Lecture Notes in Computer Science 16173

Founding Editors

Gerhard Goos
Juris Hartmanis

Editorial Board Members

Elisa Bertino, *Purdue University, West Lafayette, IN, USA*
Wen Gao, *Peking University, Beijing, China*
Bernhard Steffen, *TU Dortmund University, Dortmund, Germany*
Moti Yung, *Columbia University, New York, NY, USA*

The series Lecture Notes in Computer Science (LNCS), including its subseries Lecture Notes in Artificial Intelligence (LNAI) and Lecture Notes in Bioinformatics (LNBI), has established itself as a medium for the publication of new developments in computer science and information technology research, teaching, and education.

LNCS enjoys close cooperation with the computer science R & D community, the series counts many renowned academics among its volume editors and paper authors, and collaborates with prestigious societies. Its mission is to serve this international community by providing an invaluable service, mainly focused on the publication of conference and workshop proceedings and postproceedings. LNCS commenced publication in 1973.

Josu Doncel · Nicolas Gast · Yezekael Hayel ·
Vincenzo Mancuso
Editors

Network Games, Artificial Intelligence, Control and Optimization

12th International Conference, NETGCOOP 2025
Bilbao, Spain, October 8–10, 2025
Proceedings

Editors
Josu Doncel
University of the Basque Country
Leioa, Spain

Yezekael Hayel
Avignon University
Avignon, France

Nicolas Gast
Inria - University Grenoble Alpes
Saint-Martin-d'Hères, France

Vincenzo Mancuso
IMDEA Networks Institute
Leganés, Spain

University of Palermo
Palermo, Italy

ISSN 0302-9743 ISSN 1611-3349 (electronic)
Lecture Notes in Computer Science
ISBN 978-3-032-09314-1 ISBN 978-3-032-09315-8 (eBook)
https://doi.org/10.1007/978-3-032-09315-8

© The Editor(s) (if applicable) and The Author(s), under exclusive license to Springer Nature Switzerland AG 2026

This work is subject to copyright. All rights are solely and exclusively licensed by the Publisher, whether the whole or part of the material is concerned, specifically the rights of translation, reprinting, reuse of illustrations, recitation, broadcasting, reproduction on microfilms or in any other physical way, and transmission or information storage and retrieval, electronic adaptation, computer software, or by similar or dissimilar methodology now known or hereafter developed.
The use of general descriptive names, registered names, trademarks, service marks, etc. in this publication does not imply, even in the absence of a specific statement, that such names are exempt from the relevant protective laws and regulations and therefore free for general use.
The publisher, the authors and the editors are safe to assume that the advice and information in this book are believed to be true and accurate at the date of publication. Neither the publisher nor the authors or the editors give a warranty, expressed or implied, with respect to the material contained herein or for any errors or omissions that may have been made. The publisher remains neutral with regard to jurisdictional claims in published maps and institutional affiliations.

This Springer imprint is published by the registered company Springer Nature Switzerland AG
The registered company address is: Gewerbestrasse 11, 6330 Cham, Switzerland

If disposing of this product, please recycle the paper.

Preface

We are very glad to present the proceedings of Netgcoop 2025. The event took place in Bilbao, Spain, from October 8 to October 10, 2025. It was organized and hosted by the University of the Basque Country.

Internet communications and services are experiencing an increase in volume and diversity both in their capacity and in their demand. This comes at the cost of an increase in the complexity of their control and optimization, mainly due to the heterogeneity in architecture as well as usage. The need for new ways of effectively and fairly allocating resources belonging to a wide set of not necessarily cooperative networks to a collection of possibly competing users is urgent and is the aim of this conference.

Netgcoop 2025 covered different areas of modeling, game theory, control and optimization, with a focus on scheduling and resource allocation in 5G/6G networks, matching in graphs as well as propagation, pricing and economic models, energy and generative Artificial Intelligence.

During the conference, we had three excellent keynote talks: Patrick Loiseau (Inria) on the Price of Fairness in matroid allocation problems, Rayadurgam Srikant (University of Illinois Urbana-Champaign) on complexity bounds on Reinforcement Learning algorithms, and Rosa Lillo (UC3M, Spain, and IBidat, uc3m-Santander Big Data Institute, Spain) on applications of the network scale-up method.

The success of the conference was largely due to the technical program committee, whose members devoted much of their time and effort to provide highly qualified reviews, and also to our sponsors, which were the University of the Basque Country, Fondation Mathématique Jacques Hadamard, and BCAM-Basque Center for Applied Mathematics. We would also like to thank the staff of Springer as well as the web chair Tania Jimenez (University of Avignon) and the publicity chair Elene Anton (University of Pau). To all, we express our deepest gratitude and utter thankfulness.

In this edition, we solicited papers of at most 11 pages. We received 23 submissions, from which 16 were accepted for presentation at the conference. Each paper was single-blindly reviewed by at least three reviewers.

Finally, we would like to thank all the authors that contributed to this conference for submitting a paper.

October 2024

Josu Doncel
Nicolas Gast
Yezekae Hayel
Vincenzo Mancuso

Organization

General Chairs

Josu Doncel   University of the Basque Country, Spain
Nicolas Gast   Inria and Université Grenoble Alpes, France

Program Committee Chairs

Yezekael Hayel   University of Avignon, France
Vincenzo Mancuso   University of Palermo, Italy, and IMDEA Networks, Spain

Steering Committee

Eitan Altman   Inria, France
Tamer Basar   University of Illinois Urbana-Champaign
Tijani Chahed   Télécom SudParis, France
Yezekael Hayel   University of Avignon, France
Hélène Le Cadre   Inria, France
Bruno Tuffin   Inria, France
Quanyan Zhu   New York University, USA

Program Committee

Khushboo Agarwal   Inria, France
Tansu Alpcan   University of Melbourne, Australia
Elene Anton   University of Pau, France
Veronica Belmega   CNRS, France
Ana Busic   Inria, France
Livia Chatzieleftheriou   IMDEA Networks, Spain
Salah Eddine Elayoubi   CentraleSupélec, France
Jocelyne Elias   Università di Bologna, Italy
Dieter Fiems   Ghent University, Belgium
Luis Guijarro   Technical University of Valencia, Spain
Vasileios Karyotis   Ionian University, Greece

Lasse Leskelä	Aalto University, Finland
Patrick Maille	IMT Atlantique, France
Andrea Marin	University of Venice, Italy
Iriniel-Constantin Morarescu	University of Lorraine, France
Ariel Orda	Technion, Israel
Balakrishna Prabhu	LAAS-CNRS, France
Dominique Quadri	University of Paris-Saclay, France
Alexandre Reiffers-Masson	IMT Atlantique, France
Matteo Sereno	University of Turin, Italy
Nahum Shimkin,	Technion, Israel
Alonso Silva	Nokia Bell Labs, France
Corinne Touati	Inria, France
Sabine Wittevrongel	Ghent University, Belgium

Additional Reviewers

Takuma Adams
Kim Hammar
Jose Ramon Vidal
Vicent Pla

Contents

Stationary Models of Adversarial Coupon Collection with an Application to Moving-Target Defense 1
    George Kesidis, Takis Konstantopoulos, and Michael Zazanis

Benchmarking Machine Learning Models for QoE Estimation in Video Streaming: Accuracy, Efficiency, Confidence and Explainability 13
    Miren Nekane Bilbao, Mikel Getino-Petit, and Javier Del Ser

Performance Paradoxes in Matching Systems are not that Rare 25
    A. Busic, J. M. Fourneau, A. Lunven, and S. Li

An Anti-eavesdropping Strategy in Communication with a Group of Nodes 35
    Andrey Garnaev and Wade Trappe

Economic Analysis of DMA-Constrained Data Sharing Strategies Among Platforms 47
    Patrick Maillé and Bruno Tuffin

Achieving a Collective Target Through Incentives 57
    K. S. Ashok Krishnan, Hélène Le Cadre, and Ana Bušić

Energy-Efficient Optimization of Cooperative Spectrum Sensing Algorithms in Multi-RAT Cognitive Networks 68
    Farzam Nosrati, Antonio Scarvaglieri, Mariana Falco, Fabio Busacca, Daniele Croce, and Sergio Palazzo

Coordinated Attack Planning in Probabilistic Attack Graphs within a Sensor-Allocated Network 79
    Romaric Mofouet, Haoxiang Ma, Jie Fu, Charles Kamhoua, Gabriel Deugoue, and Arnold Kouam

Best-Response Learning in Budgeted α-Fair Kelly Mechanisms 90
    Cleque Marlain Mboulou-Moutoubi, Younes Ben Mazziane, Francesco De Pellegrini, and Eitan Altman

The Effect of Network Topology on the Equilibria of Influence-Opinion Games 100
    Yigit Ege Bayiz, Arash Amini, Radu Marculescu, and Ufuk Topcu

The Power of Stories: Narrative Priming in Networked Multi-Agent LLM
Interactions 112
  Gerrit Großmann, Larisa Ivanova, Sai Leela Poduru,
  Mohaddeseh Tabrizian, Islam Mesabah, David A. Selby,
  and Sebastian J. Vollmer

Moment Constrained Optimal Transport for Energy Demand Management
of Heterogeneous Loads 123
  Julien Cardinal, Thomas Le Corre, and Ana Bušić

Expected Extremal Reward of a Markov Decision Process 135
  Olivier Tsemogne and Yezekael Hayel

Strategic Interaction Between Queueing System and Impatient User-Base 147
  Anirban Mitra, Manu K. Gupta, and N. Hemachandra

Optimizing Stealth Infections in an SI^2R Model with Active-to-Sleep
Dynamics 167
  Mohamed Arnouss, Willie Kouam, and Yezekael Hayel

Optimizing Energy in Supervised Learning with Data Summarization:
A Comparative Study 181
  O. Haddaji, O. Brun, and B. J. Prabhu

Author Index 193

Stationary Models of Adversarial Coupon Collection with an Application to Moving-Target Defense

George Kesidis[1(✉)], Takis Konstantopoulos[2], and Michael Zazanis[3]

[1] EE and CSE, Penn State University, University Park, PA, USA
gik2@psu.edu
[2] Mathematics, University of Liverpool, Liverpool, UK
[3] Statistics, Athens University of Economics and Business (AUEB), Athens, Greece
zazanis@aueb.gr

Abstract. In cybersecurity, moving-target defense (MTD) mitigates reconnaissance attacks from botnets by periodically resetting server identities, disrupting adversarial intelligence gathering. We model this defense mechanism as a dynamic coupon collection process, where servers correspond to coupon types and botnet attacks to coupon acquisitions. The defender's intervention (removing collected coupons) represents server resets, restoring their uncompromised state. In this paper, we derive various results regarding the number of compromised servers under various stochastic attack-defense dynamics.

1 Introduction

We consider an application of moving-target defense (MTD) in a cybersecurity setting [4,5,7,11,14]. Botnets conducting reconnaissance aim to identify as many different servers as possible before launching distributed denial-of-service (DDoS) attacks, wherein they overwhelm servers, rendering them inoperable for legitimate users. To mitigate this threat, MTD periodically resets server identities (IP addresses), disrupting the botnet's intelligence-gathering process and increasing the attacker's work factor. Changing the IP address of a server may require redirection messaging to its active clients. The connection identifier of the QUIC transport protocol [8] can persist when an end-host's IP address changes. Simple bots may not be able to process redirection messages or they may not support protocols such as QUIC.

We model the foregoing interaction between attacker and defender as a dynamic coupon collection problem, where coupon types correspond to distinct servers, bots act as collectors, and acquiring a coupon represents a successful reconnaissance event. The defender periodically removes collected coupons, corresponding to identity resets that restore servers to an uncompromised state. To ensure unpredictability in the defensive strategy, reset events follow a Poisson process.

This formulation provides a probabilistic framework for analyzing the number of servers that remain uncompromised under various attack-defense strategies. We primarily study the stationary distribution of collected identities and, at times, the distribution of the number of coupons for each type. The problem connects to queueing models with "catastrophic" (or "disaster") resets [2], where accumulated service content is periodically flushed, as also observed in population dynamics [3] and some emerging applications (e.g., voiding a user cache upon teleportation in Virtual Reality). By varying assumptions on attack-defense mechanisms, we derive insights into optimal strategies for mitigating botnet reconnaissance and strengthening cyber resilience. This work establishes a mathematical model for quantitatively assessing MTD approaches, leveraging probabilistic structures to enhance adaptive defensive measures against evolving threats.

The paper is organized as follows. We first interpret some known models in terms of coupon collecting used for MTD: Sect. 2 presents the case where coupon types are reset independently, while Sect. 3 analyzes simultaneous resets of all coupon types. An essential contribution of this paper is in Sect. 4, where we consider simultaneous coupon selection in fixed-size batches with synchronous resets; here, the distribution of the number of compromised servers involves Stirling numbers of the second kind. Section 5 presents numerical examples illustrating key model behaviors. Finally, we conclude with a brief summary in Sect. 6.

2 Fully Asynchronous Resetting of Coupons

There are M coupon types, each available in unlimited supply. In the classical coupon collection problem, coupons are gathered until all types have been obtained. See [6] for sharp asymptotics related to this problem.

Here, coupons are continuously collected over time. Specifically, coupons of type i are gathered at the points of a renewal process Ξ_i. The times between successive points of Ξ_i are i.i.d. copies of the positive random variable X_i, with finite expectation $EX_i = 1/\lambda_i$. Coupon type i represents a server, while the process Ξ_i corresponds to an adversary launching attacks against that server. Each adversary maintains a buffer where the selected coupons are stored.

There is, in addition, a coupon reset process Φ_i. At the points of Φ_i, all coupons stored in the buffer of adversary i are immediately cleared. We assume that Φ_i is a Poisson process with times between successive points that are i.i.d. copies of an exponential random variable S_i with expectation $ES_i = 1/\mu < \infty$.

It is assumed that the $2M$ point processes $\Xi_i, \Phi_i, i = 1, \ldots, M$, are independent.

The dynamics of the system is implicit in the above discussion. Let

$W_i(t) :=$ number of coupons present at time t in the buffer of adversary i

$$Y_i(t) = \mathbf{1}\{W_i(t) > 0\}, \quad Y(t) := \sum_i \mathbf{1}\{W_i(t) > 0\}. \tag{1}$$

It is clear, from classical applied probability arguments, that there is a stationary process $W_i(t), t \in \mathbb{R}$, satisfying the system dynamics. We shall consider this process. Then $Y(t), t \in \mathbb{R}$, is also stationary. Note that $Y(t)$ is the number of servers that are compromised at time t. Alternatively, $Y(t)$ is the number of coupon-types collected at time t. We refer to $Y(t)$ as the *collection-type size* at time t.

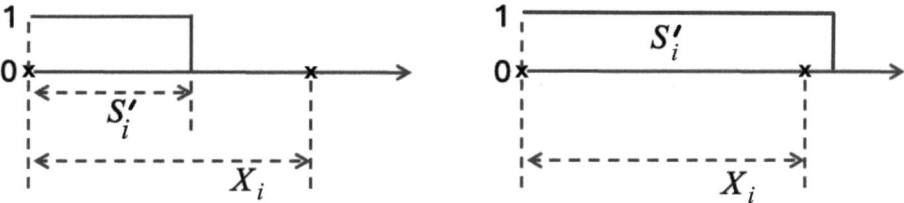

Fig. 1. $Y_i(t)$ over the interval $[0, X_i]$ for cases where $S'_i < X_i$ (left) and $S'_i > X_i$ (right).

2.1 The Distribution of the Collection-Type Size

Consider the stationary process $Y(t)$, $t \in \mathbb{R}$ and let Y denote a random variable whose distribution is the distribution of $Y(t)$ for some (and hence all) t. It may be worth noting that, for each $1 \leq i \leq M$, the $\{0,1\}$-valued process $Y_i(t)$, $t \in \mathbb{R}$, has the distribution of the buffer contents of a stationary GI/M/1/1 queue.

Proposition 1. *For all $z \in \mathbb{R}$,*

$$Ez^Y = \prod_{i=1}^{M} \frac{E\min\{S_i, X_i\}z + E(X_i - S_i)^+}{EX_i} \quad (2)$$

Proof. Fix $1 \leq i \leq M$. Let $T_i(1)$ be the first positive point of Ξ_i and S'_i the first positive point of Φ_i. Let Y_i be a random variable distributed like $Y_i(t)$ for some (and hence all) t. Let P_{Ξ_i} be the Palm probability with respect to Ξ_i and E_{Ξ_i} the expectation with respect to it. By the Palm inversion formula [1],

$$Ez^{Y_i} = \lambda_i E_{\Xi_i} \int_0^{T_i(1)} z^{Y_i(s)} ds = \lambda_i E_{\Xi_i} \int_0^{X_i} z^{Y_i(s)} ds,$$

where the second equality follows from the fact that $P_{\Xi_i}(T_i(1) = X_i) = 1$. We have

$$Y_i(s) = \mathbf{1}\{s \leq T_i(1) \wedge S'_i\}, \quad 0 \leq s \leq T_i(1).$$

Combining the above and using the fact that S'_i has the law of S_i (by virtue of the memoryless property of the exponential distribution) we obtain

$$Ez^{Y_i} = \lambda_i E(X_i z \mid S'_i > X_i) P(S'_i > X_i) + \lambda_i E(S'_i z + (X_i - S'_i) \mid S'_i < X_i) P(S'_i < X_i)$$

See Fig. 1. Since $Y = \sum_i Y_i$ is the sum of M independent random variables, we obtain (2) after a little algebra. □

Corollary 2.

$$EY = \sum_{i=1}^{M} \frac{E\min\{S_i, X_i\}}{EX_i} \quad (3)$$

Proof. We use the simple fact that $EY = \lim_{z \to 1} \frac{d}{dz} Ez^Y$ and the expression (2). □

Corollary 3. *Let*

$$c_i := \frac{E(X_i - S_i)^+}{E\min\{S_i, X_i\}}, \quad C := \prod_{i=1}^{M} \lambda_i E \min\{S_i, X_i\}.$$

Then

$$P(Y = M - k) = C \sum_{0 \leq i_1 < \cdots < i_k \leq M} c_{i_1} \cdots c_{i_k}, \quad 0 \leq k \leq M.$$

In particular, if the Ξ_i are identically distributed and the Φ_i are identically distributed then $M - Y$ is binomially distributed. The second assertion follows immediately when we take into account that the c_i are identical under the identical distribution assumptions.

Proof. Expanding the product of monomials on the right-hand side of (2) we obtain

$$Ez^Y = C \prod_{i=1}^{M} (z + c_i) = C \sum_{k=0}^{M} \left(\sum_{1 \leq i_1 < \cdots < i_k \leq M} c_{i_1} \cdots c_{i_k} \right) z^{M-k},$$

from which the probabilities can be read off directly. □

2.2 The Number of Coupons Collected of a Particular Type

We now examine the number of collected coupons for each type, motivated by scenarios involving multiple independent collectors or a single collector requiring more than one coupon per type. In the context of a security application, one may interpret this as server i being rendered inoperable once the number of attacks (i.e., collected coupons) surpasses a predefined threshold.

Let $T_i(k)$ be the kth point of Ξ_i and define

$$w_i(k) := W_i(T_i(k)+),$$

the number of coupons of type i in the buffer of adversary i just after the kth point of Ξ_i. Hence $w_i(k) \in \mathbb{N} = \{1, 2, \ldots\}$. Clearly, $w_i(k)$, $k = 1, 2, \ldots$, is Markovian. Its (nonzero) transition probabilities are

$$P(w_i(k+1) = w_i(k) + 1 \mid w_i(k)) = P(S_i > X_i) = G_i(\mu_i)$$
$$P(w_i(k+1) = 1 \mid w_i(k)) = P(S_i < X_i) = 1 - G_i(\mu_i),$$

where

$$G_i(v) := Ee^{-vX_i}.$$

It is then clear that the stationary distribution π_i of w_i is geometric.

Proposition 4. *Let w_i be a random variable whose distribution is that of $W_i(T_k+)$ for some (and hence all) k. Then*

$$P(w_i = n) = (G_i(\mu_i))^{n-1}(1 - G_i(\mu_i)), \quad n \in \mathbb{N}.$$

From this, we easily obtain by conditioning on whether $Y_i = 0$ (see Corollary 2):

Corollary 5. *Let W_i be a random variable whose law is the law of $W_i(t)$ for some (and hence all) t. Then*

$$P(W_i = n) = \pi_i(n) \frac{E \min\{X_i, S_i\}}{EX_i}, \quad n \geq 1$$

$$P(W_i = 0) = \frac{E(X_i - S_i)^+}{EX_i}$$

Thus, the distribution of W_i is a mixture of a geometric law and a unit mass at 0.

3 Fully Synchronous Resetting of Coupons

We now consider a model in which all coupons are reset simultaneously at the points of a Poisson process Φ with rate μ, thereby enhancing the effectiveness of the defense system.

To ensure tractability, we assume that coupon selection takes place at the points of a single renewal process Ξ, where coupon i is selected with probability α_i. We have

$$\sum_{i=1}^{M} \alpha_i = 1.$$

We now assume that Ξ has rate $M\lambda$.

The statement at the beginning of Sect. 2.1 regarding the independence of different coupon types no longer holds.

3.1 The Collection-Type Set

As before, we let $W_i(t)$ be the number of coupons of type i in the buffer of adversary i at time t. Let $Y(t)$ be the collection-type size at time t. In general, $Y(t), t \in \mathbb{R}$, is not Markovian. We are thus forced to consider the *collection-type set*

$$\mathsf{Y}(t) := \{i : 1 \leq i \leq M, W_i(t) > 0\} \subset \{1, \ldots, M\},$$

so that

$$Y(t) = |\mathsf{Y}(t)|.$$

Moreover, let $T(k)$ be the kth positive point of Ξ and define

$$\mathsf{y}(k) := \mathsf{Y}(T(k)+) \subset \{1, \ldots, M\} \setminus \{\varnothing\}.$$

Let X denote a random variable distributed as $T(k+1) - T(k)$ and S an independent exponential random variable with rate μ. Set

$$H(\theta) := Ee^{-\theta X}$$

for the Laplace transform of X and let

$$\alpha(\mathsf{z}) := \sum_{i \in \mathsf{z}} \alpha_i, \quad \mathsf{z} \subset \{1, \ldots, M\}.$$

We then have the following.

Proposition 6. *The random sequence* $y(k)$, $k \in \mathbb{Z}$, *is a set-valued Markov chain. Its transition probabilities are given by*

$$P(y(k+1) = y(k) \mid y(k)) = H(\mu)\alpha(y(k)), \quad \text{if } |y(k)| > 1,$$
$$P(y(k+1) = \{i\} \mid y(k) = \{i\}) = \alpha_i,$$
$$P(y(k+1) = \{i\} \mid y(k)) = (1 - H(\mu))\alpha_i, \quad \text{if } y(k) \neq \{i\},$$
$$P(y(k+1) = y(k) \cup \{i\} \mid y(k)) = H(\mu)\alpha_i, \quad \text{if } i \notin y(k),$$

for $1 \le i \le M$. *There is a unique stationary distribution* π *for this Markov chain.*

Proof (Sketch of Proof). Consider the collection-type size $y(k)$ just after $T(k)$. If $y(k)$ is not a singleton and if there is no coupon resetting (which occurs with probability $P(S > X) = H(\mu)$) between $T(k)$ and $T(k+1)$ then $y(k+1) = y(k)$ if and only of the coupon type collected at time $T(k+1)$ is an element of the set $y(k)$; the probability of the latter is $\alpha(y(k))$. This explains the first line. The remaining lines follow a similar argument. The Markovian property of $y(k)$, $k \in \mathbb{Z}$, is obvious. The uniqueness of the stationary distribution follows from the finiteness of the state space of the chain together with the irreducibility of its transition probabilities. □

We do not attempt to obtain the stationary distribution of the above Markov chain in closed form. The above proposition allows us of course to compute it numerically. However, we easily obtain the following.

Corollary 7. *If Y is a random variable distributed as $|Y(t)|$ for some (and hence all) t under stationarity, we have*

$$P(Y = j) = \sum_{|y|=j} \pi(y) \frac{E\min\{S, X\}}{EX}, \quad 1 \le j \le m \qquad (4)$$

$$P(Y = 0) = \frac{E(X - S)^+}{EX} \qquad (5)$$

The reason that
$$y(k) = |y(k)|, \quad k \in \mathbb{Z},$$
fails to be Markovian, in general, is due to asymmetry. The asymmetry vanishes when all the α_i are identical.

3.2 The Collection-Type Size in the Fully Synchronous Symmetric System

Proposition 8. *Assume that the system is fully symmetric, that is,*

$$\alpha_i = \frac{1}{M}, \quad 1 \le i \le M.$$

Then $y(k)$, $k \in \mathbb{Z}$, is Markovian. Its transition probabilities are given by

$$\begin{aligned} P_{i,i} &= H(\mu)\tfrac{i}{M} + (1 - H(\mu))\mathbf{I}\{i=1\} & i &= 1, \ldots, M \\ P_{i,i+1} &= H(\mu)\left(1 - \tfrac{i}{M}\right) & i &= 1, \ldots, M-1 \\ P_{i,1} &= 1 - H(\mu), & i &= 2, \ldots, M \end{aligned}$$

Solving the balance equations immediately yields the stationary distribution.

Proposition 9. *Under the fully symmetric conditions, let y be a random variable distributed as the collection-type size process at a fixed (and hence) any k.*

$$P(y=i) = \frac{\left(\frac{H(\mu)}{M}\right)^{i-1} \frac{(M-1)!}{(M-i)!} \prod_{j=i+1}^{M}\left(1 - H(\mu)\frac{j}{M}\right)}{\prod_{j=1}^{M-1}\left(1 - H(\mu)\frac{j}{M}\right)}, \quad i=1,\ldots,M$$

with $\prod_{j=M+1}^{M}(\ldots) = 1$.

Remark 10. We can easily pass on from the stationary collection-type size distribution $P(y=i)$ at the collection epochs to the stationary collection-type size distribution $P(Y=i)$ at an arbitrary point of time by Palm inversion as, e.g., in Corollary 7:

$$P(Y=0) = 1 - \frac{M\lambda}{\mu}(1 - H(\mu))$$

$$P(Y=i) = \frac{M\lambda}{\mu}(1 - H(\mu))P(y=i), \quad i=1,\ldots,M$$

Remark 11. If, in addition to symmetry, the collection-type size process Ξ is also Poisson (with rate $M\lambda$) then $H(\mu) = M\lambda/(M\lambda + \mu)$ and Y has identical law as y.

4 Batch Coupon Selections and Fully Synchronous Resetting

So far, we assumed that exactly one coupon is collected at each collection time, that is, at each point of Ξ. Suppose now that N coupons are collected, uniformly at random.

Assume that Ξ is a renewal process with rate λ and that Φ is an independent Poisson process with rate μ.

Since we are interested in the distribution of the collection-type size, we first compute the distribution of the size of different coupon types collected at a given collection time. Combinatorially, we have are faced with the problem of N unordered selections with repetition from M ordered items (the coupon types).

Proposition 12. *Let M, N be two unrelated positive integers. An urn contains M coupon types with labels $1, \ldots, M$. We select N items, with repetition, and without caring about the order, uniformly at random. Let ζ be the number of different coupon types in the sample, a random integer in the set $\{1, \ldots, N \wedge M\}$. Then*

$$P(\zeta = r) = \frac{S(N,r)(M)_r}{M^N} =: P_{M,N}(r), \quad 1 \leq r \leq M \wedge N, \tag{6}$$

where $(M)_r = M(M-1)\cdots(M-r+1)$ and $S(N,r)$ is the Stirling number of the second kind, that is,

$$S(N,r) = \text{ number of partitions of a set of size } N \text{ in } r \text{ parts.}$$

Moreover,

$$E\zeta = M\left[1 - (1 - \tfrac{1}{M})^N\right] =: b_{M,N} \quad (\text{with } b_{M,0} = 0). \tag{7}$$

Proof. We prove (7) first. The expectation of ζ equals M times the probability that a coupon of specific type is selected; this probability equals $1 - (1 - \frac{1}{M})^N$.

There many ways to prove (6). One way is to note that $r! S(N, r)$ is the number of surjective functions from $\{1, \ldots, N\}$ onto a set of size r. This has to multiplied by $\binom{M}{r}$– the number of subsets of $\{1, \ldots, M\}$ of size r. Multiplying these together and dividing by M^N–the total number of functions from $\{1, \ldots, N\}$ into $\{1, \ldots, M\}$–gives (6). □

If we denote explicitly the dependence of ζ on N and call it ζ_N then it is clear that ζ_N, $N = 1, 2, \ldots$ is a Markov chain with $P(\zeta_{N+1} = r | \zeta_N = r) = r/M$ and $P(\zeta_{N+1} = r + 1 | \zeta_N = r) = 1 - r/M$. Based on this, we can derive a recursion for $\phi_N(t) = Et^{\zeta_N}$:

$$\phi_{N+1}(t) = t\phi_N(t) + \frac{1}{M}t(t-1)\phi_N'(t)$$

that can, in principle, be iterated starting from $\phi_1(t) = 1$. But there is no way to get an explicit formula.

Remark 13. As a consequence of the fact that $P_{M,N}$, given by (6), is a probability distribution, we obtain that $\sum_{r \geq 1} S(N, r)(M)_r = M^N$. See also the classic paper [12] of Rota and [10]. This is known to hold even if M is replaced by a positive real number θ. So $P_{\theta,N}$ is a probability distribution for all $\theta > 0$ and $N \in \mathbb{N}$. It would thus be tempting to call $P_{\theta,N}$ a Stirling distribution of the second kind. However, the term is already taken: In, e.g., [9, §4.12.3], the term refers to the law of the sum of n independent Poisson random variables conditional on each being strictly positive.

As before, let $Y(t)$ be the collection-type size at $t \in \mathbb{R}$, and let $y(k) = Y(T(k)+)$, where $T(k)$ is the kth point of Ξ.

Proposition 14. *The random sequence $y(k)$, $k \in \mathbb{Z}$, is Markovian. Its transition probabilities are given by*

$$P(y(k+1) = j \mid y(k)) = H(\mu) Q_{M,N}(j - y(k) | y(k)) 1_{y(k) \leq j \leq M} + (1 - H(\mu)) Q_{M,N}(j|0), \tag{8}$$

where, for $y \leq M$ and $\ell \leq \min\{N, M - y\}$,

$$Q_{M,N}(\ell|y) = \frac{\sum_{i=\ell}^{N} \binom{N}{i}(1 - \frac{y}{M})^i (\frac{y}{M})^{N-i} P_{M-y,i}(\ell)}{\sum_{\ell'=0}^{N \wedge (M-y)} \sum_{i'=\ell'}^{N} \binom{N}{i'}(1 - \frac{y}{M})^{i'} (\frac{y}{M})^{N-i'} P_{M-y,i'}(\ell')}, \tag{9}$$

$$Q_{M,N}(\ell|0) = P_{M,N}(\ell).$$

The demonstration of the above is routine, provided that we observe that the quantity $Q_{M,N}(\ell|y)$ is simply the probability that ℓ new coupon types are collected in addition to the existing y.

Remark 15. The above Markov chain is irreducible in a finite set. So it has a unique stationary distribution which can, in principle, be derived from (8), (9). The algebra is too unwieldy, so we resort into a numerical solution of the equations, which is the subject of the last section. From the distribution of y we obtain the distribution of Y using (4) and (5). In particular, of Ξ is Poisson, the two distributions coincide.

Proposition 16. *For the model of this section with a renewal process Ξ, the number of coupon types collected in stationarity are given by*

$$P(Y(0) = 0) = 1 - \frac{\lambda}{\mu}[1 - H(\mu)]$$

$$P(Y(0) = r) = \sum_{k=1}^{\infty} \frac{\lambda}{\mu}[1 - H(\mu)]^2 (H(\mu))^{k-1} P_{M,kN}(r), \quad r = 1, 2, \ldots, M. \quad (10)$$

In particular, the expected number of coupon types in stationarity is

$$EY(0) = \frac{\lambda}{\mu} M[1 - H(\mu)] \frac{1 - \left(1 - \frac{1}{M}\right)^N}{1 - H(\mu)\left(1 - \frac{1}{M}\right)^N}. \quad (11)$$

Proof. Since Ξ and Φ are two independent, stationary point processes (the first a renewal process and the second a Poisson process with rate μ) at time 0 the number of different coupon types that have been collected is determined by considering the number of collection epochs (of Ξ) in the interval of time between the latest reset epoch (of Φ) prior to 0 and 0. If we denote the points of Ξ by $\{T(n)\}_{n \in \mathbb{Z}}$ and those of Φ by $\{U(n)\}_{n \in \mathbb{Z}}$ and using the standard convention that they are numbered so that, P–a.s., $T(0) \leq 0 < T(1)$ and $U(0) \leq 0 < U(1)$,

$$p_k = P(T(-k) < U(0) < T(-k+1) < 0), \quad k = 1, 2, \ldots, \quad \text{and } p_0 = P(T(0) < U(0) \leq 0).$$

By time reversibility, the independence of the two point processes, and the fact that both $-U(0)$ and $U(1)$ are exponential with rate μ and independent, we have equivalently

$$p_k = P(T(k) < U(1) < T(k+1)) = E \int_0^\infty \mathbf{1}\{T(k) < t < T(k+1)\} \mu e^{-\mu t} dt$$

$$= E \int_0^\infty \mathbf{1}\{T(k+1) > t\} \mu e^{-\mu t} dt - E \int_0^\infty \mathbf{1}\{T(k) > t\} \mu e^{-\mu t} dt$$

$$= E e^{-\mu T(k+1)} - E e^{-\mu T(k)} = \frac{\lambda}{\mu}[1 - H(\mu)](H(\mu))^{k-1} - \frac{\lambda}{\mu}[1 - H(\mu)](H(\mu))^k$$

$$= \frac{\lambda}{\mu}[1 - H(\mu)]^2 (H(\mu))^{k-1}, \quad k = 1, 2, \ldots.$$

In the above string of equalities we have also used the fact that Ξ is a *stationary* renewal process with mean time between points λ^{-1} and therefore the first point to the right of 0 has Laplace transform $Ee^{-sT(1)} = \frac{\lambda}{s}[1 - H(s)]$. Similarly,

$$p_0 = 1 - \frac{\lambda}{\mu}[1 - H(\mu)].$$

Note that, on the event $\{T(-k) < U(0) < T(-k+1) < T(0) < 0\}$, there are k coupon collection epochs following the last reset and hence the probability of collecting r different types for coupons is given by $P_{M,kN}(r)$ as given in (6). These considerations establish (10).

Arguing similarly and using (7) we obtain

$$EY(0) = \sum_{k=1}^{\infty} p_k b_{M,Nk} = \sum_{k=1}^{\infty} \frac{\lambda}{\mu}[1 - H(\mu)]^2 (H(\mu))^{k-1} M \left[1 - \left(1 - \frac{1}{M}\right)^{Nk}\right].$$

Evaluating the above geometric series establishes (11). □

Remark 17. When the process Ξ is Poisson (with rate λ) the expression in (11) becomes
$$EY(0) = \frac{M\lambda \left(1 - \left(1 - \frac{1}{M}\right)^N\right)}{\lambda + \mu - \lambda \left(1 - \frac{1}{M}\right)^N}.$$

Remark 18. Suppose we take $N \to \infty$. In this extreme case, all M types of coupons are collected at every collection point of Ξ. That is, in the proof, $b_{M,iN} = M$ for all $1 \leq i \leq k$ (and still $b_{M,0} = 0$). So, the Palm inversion formula gives
$$EY = \mu \left[0 \cdot E\min\{X_1, S\} + M \cdot E(S - X_1)^+\right] = \mu M \frac{\lambda}{\mu(\mu + \lambda)} = M \frac{\lambda}{\mu + \lambda},$$
with $S \sim \exp(\mu)$ and $X_1 \sim \exp(\lambda)$ independent. This is equal to the expression for EY in the statement of Proposition 16 after taking $N \to \infty$.

5 Numerical Example

We conducted numerical experiments to evaluate various specific scenarios related to the cases described above. Figure 2 illustrates a representative example of a hybrid scenario in which, at coupon-type reset epochs, each coupon type is independently reset with probability q at the ticks of a Poisson process (Remark 15). Simultaneously, the collector selects fixed-size batches of coupons (N) at a constant rate (Sect. 4)[1].

In Fig. 2, the collector selects batches of $N = 10$ coupons at a fixed rate of $\lambda = 1$ batch selections per second ($\lambda = 1/EX$). The system comprises $M = 100$ coupon types, with each type reset independently with probability $q = 0.01$ at the ticks of a Poisson process with mean rate $\mu = 1/ES$, which varies along the x-axis of the figure. Consequently, the total mean coupon selection rate is $\lambda N = 10$ coupons per second, while the mean coupon-type reset rate is $q\mu = \mu/10$ types per second.

The figure reveals a "phase transition" interval in which the coupon-type reset process must exceed a certain threshold rate to successfully "defeat" the collector. A similar phase transition is observed when $q\mu$ is held constant while varying the collection rate $N\lambda$.

Finally, consider the model presented in Sect. 2, i.e., individual random coupon selections occurring at the ticks of a renewal process with rate λ, alongside fully asynchronous resets of the M coupon types at the ticks of independent Poisson processes with rate μ/M. If we take the special case where coupon selection follows a Poisson process, the problem reduces to an independent M/M/1/1 queue for each coupon type. In this setting, the expected number of collected coupons is $EY = M(\lambda/M)/(\lambda/M + \mu/M) = M\lambda/(\lambda + \mu)$, implying that EY is simply a convex function of μ. Thus, unlike Fig. 2, no phase transition is observed in this case.

---

[1] See [13] for an approximation of $S(N, r)$ as $N \to \infty$ that is uniformly accurate over r.

6 Summary

In summary, we analyzed various models of coupon collection and coupon-type resetting, focusing on the stationary distribution—particularly the stationary mean number of collected coupon types and the stationary mean number of coupons collected per type. The coupon-type resetting (clearance) processes were Poisson, and independent of each other and of the renewal collection process. Our models incorporated both synchronous (batch) and asynchronous mechanisms for coupon collection and coupon-type resetting. The main contribution was analysis of a case with batch coupon collection involving a Stirling distribution of the second kind. We concluded with a numerical example illustrating batch coupon collection, which involved a Stirling distribution of the second kind. Some of these models exhibited a "phase transition" in the expected number of collected coupons as the mean collection rate to reset rate decreased.

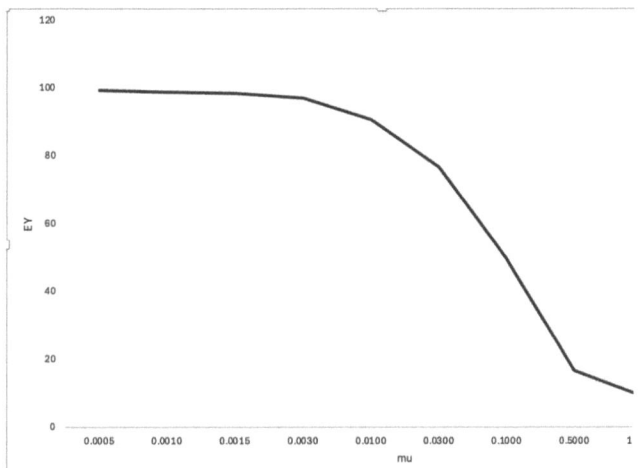

Fig. 2. Typical simulation result showing the stationary mean (EY) of the number of collected coupon types (Y). Here, the collector process has fixed rate $\lambda = 1$ batches/s where each batch size is of $N = 10$. Each coupon type is randomly reset with probability $q = 0.1$ at the ticks of a Poisson process with mean rate μ (x-axis). The Stirling distribution (6) was sampled and the total simulation time was 10^6 s.

References

1. Baccelli, F., Brémaud, P.: *Elements of Queueing Theory*. In: Springer-Verlag, Application of Mathematics: Stochastic Modelling and Applied Probability, No. 26, New York, NY (1991)
2. Boxma, O.J., Perry, D., Stadje, W.: Clearing models for M/G/1 queues. Queueing Syst. **38**, 287–306 (2001)
3. Brockwell, P.J., Gani, J., Resnick, S.I.: Birth, immigration and catastrophe processes. Adv. Appl. Prob. **14**, 709–731 (1982)

4. Carroll, T.E., Crouse, M., Fulp, E.W., Berenhaut, K.S.: Analysis of Network Address Shuffling as a Moving Target Defense. In Proc, IEEE ICC (2014)
5. Crouse, M., Prosser, B., Fulp, E.W.: Probabilistic performance analysis of moving target and deception reconnaissance defenses. In: *Proc. ACM Workshop on Moving Target Defense (MTD)*, pp. 21–29 (2015)
6. Erdős, P., Rényi, A.: On a classical problem of probability theory. Acta Math. Acad. Sci. Hung. **12**, 261–267 (1961)
7. Fleck, D., Kesidis, G., Konstantopoulos, T., Nasiriani, N., Shan, Y., Stavrou, A.: Moving-target defense against botnet reconnaissance and an adversarial coupon-collection model. In: *Proc. IEEE Conf. Dependable & Secure Computing (DSC)*, pp. 1–8 (2018)
8. Iyengar, J., Thomson, M.: QUIC: A UDP-Based Multiplexed and Secure Transport. https://datatracker.ietf.org/doc/rfc9000/. Accessed 19 Feb 2022
9. Johnson, N.L., Kemp, A.W., Kotz, S.: *Univariate Discrete Distributons*. Wiley, 3rd Edition (2005)
10. Kesidis, G., Konstantopoulos, T., Zazanis, M.A.: The generating functions of Stirling numbers of the second kind derived probabilistically. Math. Sci. **43**(2), 82–87 (2018)
11. Kesidis, G., Shan, Y., Fleck, D., Stavrou, A., Konstantopoulos, T.: An adversarial coupon-collector model of asynchronous moving-target defense against botnet reconnaissance. In: *Proc. Intern. Conf. on Malicious and Unwanted Software (MALWARE)*, pp. 61–67 (2018)
12. Rota, G.-C.: The number of partitions of a set. Amer. Math. Monthly **71**(5), 498–504 (1964)
13. Temme, N.M.: Asymptotic estimates of Stirling numbers. Stud. Appl. Math. **89**(3), 233–243 (1993)
14. Venkatesan, S., Albanese, M., Amin, K., Jajodia, S., Wright, M.: A moving target defense approach to mitigate DDoS attacks against proxy-based architectures. In: *Proc. IEEE Conf. Comm. Netw. Security (CNS)*, pp. 198–206 (2016)

Benchmarking Machine Learning Models for QoE Estimation in Video Streaming: Accuracy, Efficiency, Confidence and Explainability

Miren Nekane Bilbao[1]([✉]), Mikel Getino-Petit[1], and Javier Del Ser[1,2]

[1] University of the Basque Country UPV/EHU, 48013 Bilbao, Spain
nekane.bilbao@ehu.eus
[2] TECNALIA, Basque Research and Technology Alliance (BRTA), 48160 Derio, Spain
javier.delser@tecnalia.com

Abstract. The accurate prediction of Quality of Experience (QoE) in video streaming services is essential for optimizing user satisfaction and network performance. While traditional Quality of Service (QoS) metrics provide objective measurements of network behavior, they often fail to reflect the subjective nature of user experience. This paper investigates the use of Machine Learning models to estimate QoE based on QoS indicators. Building upon the recently published SNESet dataset, we evaluate a range of modern regression techniques, including randomization-based neural networks, symbolic regression and Kolmogorov-Arnold Networks, alongside other traditional and ensemble-based models. A central focus of this study is the explainability of such new models, which enables the extraction of domain-relevant insights from the learned relationships. Using model-agnostic techniques for explainable Artificial Intelligence and uncertainty quantification, we assess the confidence of such models in their predictions and analyze the contribution of individual features to the estimated QoE. Our results underscore the need for explainable QoE prediction systems, closing the gap between data-driven modeling and domain expertise.

Keywords: Quality of Experience · Video Streaming · Machine Learning · Conformal Prediction · Explainable AI

1 Introduction

The proliferation of video streaming platforms has made QoE a critical concern for service providers in their aim to ensure user satisfaction and retention [5]. While QoS metrics (e.g., latency, jitter, and packet loss) can provide objective insights into network performance, they often fail to capture the subjective nature of user experience when consuming such services [2]. Overcoming this weakness requires models that can characterize complex, nonlinear relationships between QoS indicators and user-perceived QoE [3].

In this context, recent advances in Machine Learning (ML) have enabled the development of data-driven models capable of estimating QoE from QoS data with increasing accuracy [11]. However, many of these models, particularly deep learning architectures, operate as "black boxes", limiting their utility in contexts where transparency and trust are essential [8]. This limitation is of utmost importance in manifold use cases, such as adaptive bit-rate streaming platforms (e.g., Netflix, YouTube), live event broadcasting (e.g., sports or concerts), and mission-critical applications like telemedicine or remote education, where understanding the rationale behind QoE predictions is vital for real-time service assurance and proactive resource allocation policies. In such scenarios, techniques and tools for explainable Artificial Intelligence (XAI) [1,4] can identify which QoS parameters most influence user experience, enabling network operators to take targeted actions. Moreover, uncertainty quantification (UQ) becomes essential when predictions are used to trigger automated control mechanisms or inform service-level agreements (SLAs), as overconfident or poorly calibrated models can lead to suboptimal resource allocation policies and ultimately, to degraded user satisfaction. Therefore, integrating explainability and confidence estimation into ML-based QoE prediction models is not only a matter of academic interest, but also a practical necessity for robust real-world video streaming systems.

This study aligns with this rationale by exploring modern ML approaches not only to predict QoE effectively, but also to provide insights into the underlying mechanisms of their prediction process. Using the SNESet dataset [11], we evaluate a range of models, from traditional regression methods to advanced models that have not been yet explored for this particular application, including randomization-based neural networks [20], Kolmogorov-Arnold Networks (KANs) [12], and symbolic regression [7]. A key contribution of this work is the interpretation of the knowledge encoded by these models through the exploration of their inherently interpretable structure or by the use of model-agnostic XAI techniques. These interpretations can be linked to domain-specific knowledge, allowing for the validation of the relevance of QoS metrics known to influence the user-perceived QoE. This alignment of the models' outputs with expert knowledge and the quantification of the models' confidence in their predictions not only enhance trust, but also facilitate their integration into operational workflows for network optimization and user experience management.

The rest of the manuscript is structured as follows: first, Sect. 2 revisits relevant references on the use of ML for QoE estimation, and clearly poses the contribution of the work w.r.t. the reviewed literature. Next, Sect. 3 details the ML models and XAI/UQ techniques under consideration. Section 4 presents the experimental setup and discusses the results obtained therefrom. Finally, Sect. 5 concludes the paper by summarizing findings and future research directions.

2 Related Work and Contribution

The relationship between QoS and QoE has been extensively studied in the context of multimedia services [19]. Traditional approaches often rely on parametric models that attempt to map specific QoS metrics (e.g., latency, jitter, and packet loss) to subjective QoE scores, as outlined in foundational works and ITU recommendations [3,10]. However, these models typically assume linear or fixed functional relationships, which fail to capture the complex, nonlinear dynamics of real-world network conditions and user behavior.

Recent research has shifted toward data-driven methods, leveraging ML to model the correlation between QoS and QoE related metrics [2]. A comprehensive review by [23] underscores the increasing adoption of predictive and proactive QoE management techniques, emphasizing the potential of ML to address the limitations of reactive or hand-crafted models. Similarly, [25] provides an overview of the challenges in QoE prediction for adaptive video streaming, noting the influence of factors such as initial delay, stalling, and quality adaptation, and the limitations of existing datasets and models in reflecting real-time user experience. In this latter regard, [11] has recently introduced the SNESet dataset and benchmarked several supervised learning regression models for QoE prediction in edge-hosted video streaming scenarios. Other studies have explored different ML-based QoE estimation pipelines, including Bayesian networks [22,24], recurrent neural networks [9], and ensemble methods [6], highlighting the growing variety of ML approaches in this domain.

Contribution. Despite their predictive power, most ML models used to date for QoE estimation lack interpretability, limiting their applicability in real-world video streaming networks. To address this niche, this work systematically evaluates the performance and complexity of both interpretable and non-interpretable models for QoE prediction, with a particular focus on their transparency, reliability and trust through the use of XAI tools and UQ techniques.

3 Machine Learning Models and UQ/XAI Techniques

We proceed by detailing the ML models (Subsect. 3.1), as well as the UQ and XAI techniques (Subsect. 3.2) under consideration.

3.1 Machine Learning Models

To begin with, we consider the wide family of Random Vector Functional Link (RVFL) networks, which are a class of single-layer feedforward randomization-based neural networks that enhance traditional architectures by incorporating both linear and nonlinear transformations of the input [16]. In RVFL, the weights between the input and hidden layer are randomly assigned and fixed, while only the output weights are learned via a closed-form solution. This randomization-based learning procedure significantly reduces training time and computational complexity. Despite their simplicity, RVFL networks can approximate complex functions effectively and are particularly attractive for real-time or resource-constrained applications. Ensemble Deep RVFL (edRVFL) [18] extends the RVFL framework by stacking multiple RVFL blocks in a deep architecture and combining their outputs. Each block consists of randomly initialized hidden layers, and skip connections are used to propagate information across layers. An ensemble strategy aggregates predictions from all blocks [15].

The second ML model under consideration is KAN, a recently proposed neural network architecture designed for function approximation, with a strong emphasis on interpretability [12]. Unlike traditional neural networks that apply fixed activation functions at each node, KANs place learnable univariate functions (typically modeled as B-splines) on the edges between nodes. KANs can model complex nonlinear relationships while maintaining a structure that is more transparent and easier to analyze than their traditional counterparts. The use of interpretable spline functions allows visualizing and understanding how each input feature contributes to the output, making KANs particularly suitable for applications where model explainability is critical.

Finally, in terms of modeling our experiments will consider symbolic regression (SR) [7], which searches for mathematical expressions that best approximate a given modeling dataset without assuming a predefined model structure. SR explores a vast space of possible equations, combining mathematical operators, constants, and input variables to discover both the structure and parameters of the model. This approach yields models that are not only accurate but also inherently interpretable, as the resulting expressions can be directly analyzed and understood by humans. To perform SR, one effective method is Genetic Programming (GP) [14], a type of evolutionary algorithm that evolves populations of candidate expressions over successive generations. Through search operators, GP iteratively refines symbolic expressions to minimize prediction error while balancing model complexity, ultimately producing compact and explainable regression models that reveal meaningful relationships from the modeled data.

3.2 UQ and XAI Techniques

To complement the predictive capabilities of the aforementioned ML models, we consider two techniques (SHAP [13] and Conformal Prediction, CP [17]). While models such as SR and KANs offer inherent interpretability through explicit mathematical expressions, others like edRVFL are considered black-box models. In such cases, SHAP provides a principled framework for attributing the contribution of each input feature to individual predictions. Grounded in cooperative game theory, SHAP enables both local and global interpretability through visual tools such as waterfall and summary plots, thereby enhancing transparency and trust in model outputs. In parallel, CP serves as a complementary tool that quantifies the uncertainty surrounding each prediction issued by the model. By constructing statistically valid prediction intervals under the assumption of data exchangeability, CP provides formal guarantees on coverage, independent of the underlying model. This is valuable in the context of video streaming management, where understanding the predicted QoE metrics and the confidence in such predictions can inform adaptive resource allocation strategies.

4 Experiments and Results

To assess the performance, efficiency, confidence and explanations of the above models, we design experiments to answer three Research Questions (RQs):

- RQ1: How do the selected ML models perform in the QoE estimation task compared to previously established approaches in the literature?
- RQ2: To what extent does the application of CP produce statistically valid and reliable prediction intervals across different ML models?
- RQ3: Can explainability techniques like SHAP generate explanations that align with domain-specific knowledge in QoE for video streaming?

To this end, our experiments consider the following dataset and methods:

Dataset. The SNESet dataset is a comprehensive collection of network performance metrics designed to support the prediction of Quality of Experience (QoE) in video streaming services. SNESet comprises 9 million traces from 8 video streaming applications, captured during four months of year 2022 at ca. 800 localizations and 3 Internet providers. It includes a wide range of Quality of Service (QoS) indicators such as TCP connection times, ICMP round-trip times, packet loss rates, CPU and memory utilization, and various throughput-related features. These variables are the predictors input to the ML models under comparison. The target variable (QoE) is quantified using the buffer rate, defined as the proportion of playback time affected by stalling events.

Baselines. The original SNESet publication [11] benchmarked several ML models for predicting QoE from QoS metrics in video streaming. The baseline models included ElasticNet, a regularized linear regression method combining L1 and L2 penalties for feature selection and robustness; Support Vector Regression (SVR), which optimizes a margin-based loss function to handle nonlinear relationships; and Random Forest (RF), an ensemble of decision trees that improves generalization through bootstrap aggregation. Additionally, the study evaluated several boosting ensembles (XGBoost, LightGBM, and CatBoost) known for their high predictive performance and scalability. Finally, a Deep Neural Network (DNN) architecture was implemented, combining embedding layers for categorical inputs with fully connected layers for numerical features.

Evaluation Protocol and Metrics. The dataset was split using 5-fold cross-validation (CV), ensuring that each model was trained and validated on multiple partitions of the data to reduce variance and improve generalization. For each fold, the model was trained on 80% of the data and validated on the remaining 20%, rotating the validation set across folds. To optimize model performance, a randomized hyperparameter search was applied within the CV loop, using 100 iterations per model. This approach allowed efficient exploration of the hyperparameter space while avoiding the computational cost of exhaustive grid search.

To evaluate the performance of the ML models, several metrics were employed. The Mean Absolute Error (MAE) and Symmetric Mean Absolute Percentage Error (SMAPE) quantify the average magnitude of prediction errors:

$$\text{MAE} = \frac{1}{n}\sum_{i=1}^{n}|y_i - \hat{y}_i|, \quad \text{SMAPE} = \frac{100}{n}\sum_{i=1}^{n}\frac{|y_i - \hat{y}_i|}{(|y_i| + |\hat{y}_i|)/2}, \quad (1)$$

where y_i is the true value and $\hat{y}_i$ is the predicted value. To assess the reliability of uncertainty estimates, the *conformal score* for each prediction measures the deviation from the interval. It is defined as $s_i = \max\left(y_i - \hat{y}_i^U, \hat{y}_i^L - y_i, 0\right)$, where $\hat{y}_i^L$ and $\hat{y}_i^U$ are the lower and upper bounds of the prediction interval. The conformal score captures how far the true value lies outside the predicted interval, if at all. To assess predictive uncertainty, both split CP and Cross-CP methods were tested, with a target coverage level of 90%.

Implementation. The entire modeling pipeline was implemented in Python. CP was applied by using the MAPIE library [21]. All experiments were executed locally on a personal computer equipped with an Intel Core i5-1235U processor (10 cores at 1.3GHz), 8 GB DDR4 RAM, and no dedicated GPU. Code and results are available at https://github.com/javierdelser/QoE_XAI_UQ.

4.1 RQ1: Performance Versus Computational Efficiency

We start by addressing RQ1 based on the analysis of MAE and SMAPE results across different models and dataset sizes. The scores reported in Table 1 (best results highlighted as ▇) reveal that traditional models such as ElasticNet and SVR exhibit relatively stable but higher error rates across all dataset sizes, with MAE values consistently above 2.2 and SMAPE exceeding 60%, indicating limited capacity to capture the complex QoS-QoE relationships. Ensemble models like LightGBM and CatBoost show improved performance, particularly on larger datasets, with MAE values dropping below 2.2 and SMAPE around 60%, reflecting their ability to model non-linear interactions more effectively. Deep learning models such as DNN further reduce the error, especially on larger datasets, but at the cost of significantly higher computational demands. Among the advanced interpretable models, edRVFL consistently achieves the lowest MAE (as low as 1.63) and SMAPE (49%), demonstrating high accuracy and robustness across dataset sizes. KAN also performs competitively on smaller datasets, while SR using GP (SR-GP) offers a good trade-off between interpretability and accuracy.

Our response to RQ1 follows by inspecting the training and inference latencies across models and dataset sizes. As can be observed in Table 2, traditional models like ElasticNet and SVR exhibit low inference times (typically under 0.05 s), but their training times increase significantly with dataset size. ElasticNet, for instance, grows from 2.45 s on 5K samples to nearly 470 s on 100K. Ensemble models such as LightGBM and CatBoost show moderate training times (e.g.,

Table 1. MAE and SMAPE results for different models and dataset sizes (K: $\times 10^3$).

	5K		10K		50K		100K	
Model	MAE	SMAPE	MAE	SMAPE	MAE	SMAPE	MAE	SMAPE
ElasticNet	2.3124	63.30%	2.3346	63.41%	2.3175	63.47%	2.3339	63.75%
SVR	2.2345	71.27%	2.5799	64.76%	2.3310	70.16%	2.5708	67.25%
RF	2.2211	60.16%	2.2430	60.84%	-	-	-	-
LightGBM	2.0984	60.36%	2.1854	62.26%	2.2260	62.79%	-	-
XGBoost	2.3237	61.47%	2.3430	62.29%	2.3417	62.65%	-	-
CatBoost	2.2146	60.99%	2.2681	62.18%	2.3283	63.37%	2.3032	62.94%
DNN	2.0229	58.37%	2.1757	61.82%	2.1984	63.05%	2.1681	63.01%
edRVFL	1.6310	48.76%	1.6616	49.61%	1.6602	49.56%	1.6610	49.63%
KAN	1.6096	47.66%	-	-	-	-	-	-
SR-GP	1.7433	51.09%	1.7703	51.81%	1.7600	52.64%	-	-

* Values marked with "-" correspond to models whose training did not complete within 8 hours.

between 65 and 1405 s for LightGBM) and slightly higher inference times, especially on larger datasets, due to their more complex tree structures. Deep learning models like DNN and KAN incur the highest training costs, with KAN reaching over 28,000 s on 5K samples and DNN exceeding 10,000 s on 100K, making them less practical for real-time or resource-constrained environments. In contrast, edRVFL stands out for its exceptional efficiency, with training times under 8 s and inference times under 1.5 s even on the largest dataset tested. This makes edRVFL not only the most accurate model but also the most computationally efficient, reinforcing its suitability for scalable and responsive QoE prediction systems.

Table 2. Training and inference times (seconds) for different models and dataset sizes.

	5K		10K		50K		100K	
Model	Training	Inference	Training	Inference	Training	Inference	Training	Inference
ElasticNet	2.45	0.01	8.80	0.02	208.75	0.02	469.95	0.05
SVR	5.50	0.05	177.50	0.05	287.90	0.05	704.20	0.05
RF	790.85	0.03	3,507.95	0.05	> 8h	> 8h	> 8h	> 8h
LightGBM	65.55	0.20	130.90	0.35	503.45	3.70	1,405.30	4.11
XGBoost	776.55	0.20	909.75	0.20	1,813.95	1.33	> 8h	> 8h
CatBoost	76.45	0.02	90.78	0.05	4,510.31	0.50	4,945.08	0.65
DNN	248.55	0.20	483.95	0.40	2,319.45	1.95	10,609.45	11.75
edRVFL	2.05	0.21	1.95	0.35	4.10	0.60	7.40	1.45
KAN	28,719.20	4.75	> 8h	> 8h	> 8h	> 8h	> 8h	> 8h
SR-GP	2,712.85	0.03	3,138.65	0.05	9062.15	0.65	> 8h	> 8h

4.2 RQ2: Differences Between Conformalized Prediction Intervals

To address RQ2, we analyze the conformal scores reported in Table 3. These scores quantify the deviation of true values from the predicted intervals, with lower (closer to zero) values indicating tighter and more reliable coverage. Across all dataset sizes, edRVFL consistently achieves the lowest conformal scores (e.g., −0.01105 to −0.01344), suggesting that their prediction intervals are both narrow and well-calibrated. This indicates a high degree of reliability in the uncertainty estimates produced by CP when applied to these models. In contrast, LightGBM exhibits higher conformal scores, particularly on smaller datasets (e.g., −0.15677 on 10k), though its performance improves significantly as the dataset size increases (e.g., −0.01283 on 100k), reflecting better calibration with more data. KAN and SR-GP also show reasonably low scores on the datasets where they were evaluated, though their results are limited to smaller sample sizes. Together with the visualization of the prediction intervals produced by edRVFL for different samples (Fig. 1), these findings confirm that CP can produce statistically valid and reliable prediction intervals across a variety of ML models, with the best performance observed in models like edRVFL that combine strong predictive accuracy with stable uncertainty estimation.

Table 3. Conformal scores across models and dataset sizes.

Model	5K	10K	50K	100K
LightGBM	-0.13152	-0.15677	-0.13116	-0.01283
edRVFL	-0.01344	-0.01116	-0.01105	-0.01153
KAN	-0.01908	–	–	–
SR-GP	-0.01547	-0.01686	-0.01781	–

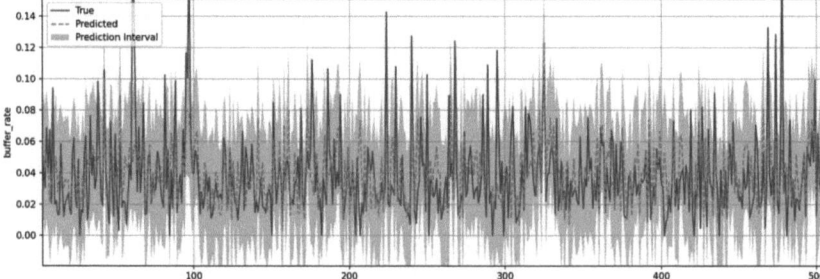

Fig. 1. Actual QoE values, predictions and confidence intervals produced by CP for several test instances and the edRVFL model.

4.3 RQ3: Alignment of Explanations with Domain Knowledge

We apply SHAP to analyze feature contributions within edRVFL to buffer rate predictions across varying dataset sizes. To this end we resort to waterfall plots, which decompose a prediction of the model into the additive contributions of each input feature, rendering a transparent view of how edRVFL produces a given buffer rate estimate. The plot begins with the model's expected value, namely, the mean prediction across the training data, and sequentially adds or subtracts SHAP values for each feature, ordered by their impact. Features that increase the prediction are shown in red, while those that decrease it appear in blue. The final bar represents the model's QoE estimation for the given instance.

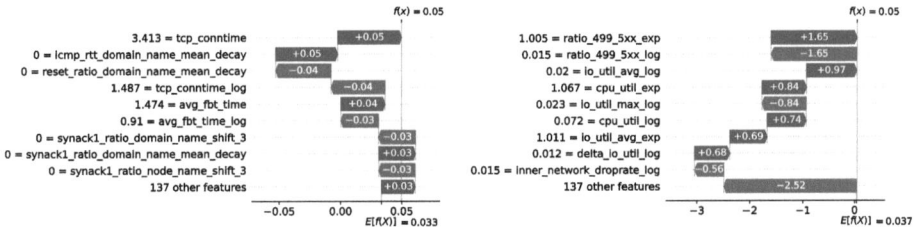

Fig. 2. SHAP waterfall plots of edRVFL for 5K (left) and 50K (right) training samples.

The waterfall plots in Fig. 2 reveal that, for smaller datasets (5K, left plot), features such as tcp_conntime, icmp_rtt, and reset_ratio are the most influential. These features are well-established indicators of network performance: tcp_conntime reflects connection setup latency, icmp_rtt captures round-trip delay, and reset_ratio signals connection instability. Their saliency in small datasets suggests that edRVFL initially relies on coarse-grained, high-impact indicators that are strongly correlated with QoE degradation. However, as the dataset size increases (50K, right plot), features related to throughput variability, packet retransmissions, and temporal aggregates (e.g., exponentially weighted moving averages) gain importance. With larger datasets, edRVFL can learn subtler interactions in network-wide metrics and the occupancy level of local computation resources that are less apparent in smaller samples. Consequently, explanations of edRVFL become more aligned with the complex nature of QoE in real-world streaming environments.

To complement the local perspective provided by the waterfall plot, we turn our focus to the SHAP summary plots in Fig. 3 to gain a global understanding of edRVFL's behavior. Unlike the waterfall plot, which explains a single prediction, the summary plot aggregates SHAP values across all predictions, showing how each feature influences the model's output throughout the dataset. Each point represents a SHAP value for a feature in one instance, with its position on the horizontal axis indicating the magnitude and direction of the feature's impact. The color of the point reflects the actual value of the feature (red for high values and blue for low values), identifying which features are

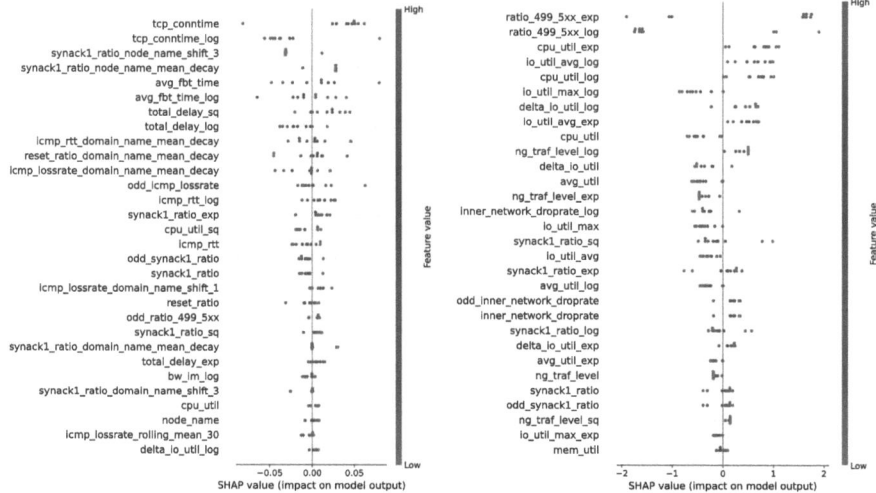

Fig. 3. SHAP summary plots of edRVFL for 5K (left) and 50K (right) training samples. (Color figure online)

important and how their values affect the prediction. Features like `tcp_conntime` and `synack1_ratio_node_name_shift_3` show a clear correlation between their magnitude and the sign of their SHAP value. High values of `tcp_conntime` (red points) tend to increase the predicted buffer rate, while high values of `reset_ratio_domain_name_mean_decay` and `icmp_rtt_domain_name_mean_decay` are associated with negative SHAP values, indicating a reduction in the prediction. As observed in the waterfall plots of Fig. 2, increasing the dataset size from 5K to 50K leads to a shift in feature importance. With more data, edRVFL begins to rely more heavily on features related to local resources (e.g. `cpu_util_log`, `cpu_util_exp`) and network-level aggregates.

5 Conclusions and Future Research

This study has evaluated and compared a range of both traditional and advanced machine learning techniques to model the complex, non-linear relationship between QoS and QoE in video streaming networks. The ultimate goal is to identify models that not only deliver high predictive accuracy, but also offer interpretability and computational efficiency, making them suitable for real-world deployment in resource-constrained network environments. To this end, three advanced modeling approaches have been considered: i) KAN, selected for its balance between expressiveness and interpretability; ii) edRVFL, due to its lightweight architecture and ensemble-based deep structure, which enables fast training and inference without backpropagation; and iii) SR-GP, for its capacity to generate explicit mathematical expressions that describe the QoS-QoE mapping, enabling straightforward human interpretability. To further enrich the analysis, two model-agnostic tools for uncertainty estimation and explainability were employed: i) SHAP, to provide both local and global interpretability; and ii) conformal prediction, to quantify predictive uncertainty through the generation of statistical prediction intervals.

Our experimental results have demonstrated that edRVFL significantly outperforms traditional approaches in terms of both accuracy (lowest MAE and SMAPE) and training time, particularly on larger datasets. Additionally, the use of SHAP and CP has enabled a deeper understanding of feature importance and uncertainty quantification, respectively, reinforcing the trustworthiness of the proposed models. The experiments have also exposed computational limitations constraining the scalability of some models on large datasets.

Future research will explore the integration of nested cross-validation to further enhance the robustness of model evaluation and hyperparameter tuning. Additionally, expanding the dataset to include more diverse network conditions and user contexts would improve generalizability. Investigating hybrid models that combine the interpretability of symbolic regression with the predictive power of deep learning could also be a promising direction to pursue. Finally, deploying the best-performing models in real-time streaming environments will provide valuable feedback on their utility for adaptive QoE optimization systems.

Acknowledgements. This research was supported by the Basque Government through research groups NQaS (IT1635-22) and MATHMODE (IT1456-22).

References

1. Ali, S., et al.: Explainable Artificial Intelligence (XAI): what we know and what is left to attain trustworthy Artificial Intelligence. Info. Fus. **99**, 101805 (2023)
2. Alreshoodi, M., Woods, J.: Survey on QoE-QoS correlation models for multimedia services. Int. J. Dist. Parall. Syst. **4**(3), 53 (2013)
3. Banović-Ćurguz, N., Ilišević, D.: Mapping of QoS/QoE in 5G networks. In: International Convention on Information and Communication Technology, Electronics and Microelectronics (MIPRO), pp. 404–408. IEEE (2019)
4. Barredo Arrieta, A., et al.: Explainable Artificial Intelligence (XAI): concepts, taxonomies, opportunities and challenges toward responsible AI. Info. Fus. **58**, 82–115 (2020)
5. Bouraqia, K., et al.: Quality of experience for streaming services: measurements, challenges and insights. IEEE Access **8**, 13341–13361 (2020)
6. Casas, P., et al.: Enhancing machine learning based QoE prediction by ensemble models. In: IEEE International Conference on Distributed Computing Systems (ICDCS), pp. 1642–1647 (2018)
7. Dong, J., Zhong, J.: Recent advances in symbolic regression. ACM Comput. Surv. **57**(11), 1–37 (2025)
8. Díaz-Rodríguez, N., Del Ser, J., et al.: Connecting the dots in trustworthy Artificial Intelligence: from AI principles, ethics, and key requirements to responsible AI systems and regulation. Info. Fusion **99**, 101896 (2023)

9. Eswara, N., et al.: Streaming video QoE modeling and prediction: a long short-term memory approach. IEEE Trans. Circuits Syst. Video Technol. **30**(3), 661–673 (2019)
10. ITU-T: P.10/G.100: Vocabulary for performance, quality of service and quality of experience (2017). https://www.itu.int/rec/T-REC-P.10-201711-I/en
11. Li, Y., et al.: Demystifying the QoS and QoE of edge-hosted video streaming applications in the wild with SNESet. Proc. ACM Manag. Data **1**(4), 1–29 (2023)
12. Liu, Z., et al.: KAN: Kolmogorov-Arnold networks. arXiv:2404.19756 (2024)
13. Lundberg, S.M., Lee, S.I.: A unified approach to interpreting model predictions. Adv. Neural Info. Process. Syst. **30** (2017)
14. Makke, N., Chawla, S.: Interpretable scientific discovery with symbolic regression: a review. Artif. Intell. Rev. **57**(1), 2 (2024)
15. Malik, A.K., et al.: Random vector functional link network: recent developments, applications, and future directions. Appl. Soft Comput. **143**, 110377 (2023)
16. Pao, Y.H., Phillips, S.M., Sobajic, D.J.: Neural-net computing and the intelligent control of systems. Int. J. Control **56**(2), 263–289 (1992)
17. Shafer, G., Vovk, V.: A tutorial on conformal prediction. J. Mach. Learn. Res. **9**(3) (2008)
18. Shi, Q., et al.: Random vector functional link neural network based ensemble deep learning. Pattern Recogn. **117**, 107978 (2021)
19. Skorin-Kapov, L., et al.: A survey of emerging concepts and challenges for QoE management of multimedia services. ACM Trans. Multimed. Comput. Commun. Appl. **14**(2s), 1–29 (2018)
20. Suganthan, P.N., Katuwal, R.: On the origins of randomization-based feedforward neural networks. Appl. Soft Comput. **105**, 107239 (2021)
21. Taquet, V., et al.: MAPIE: an open-source library for distribution-free uncertainty quantification. arXiv:2207.12274 (2022)
22. Tasaka, S.: Bayesian hierarchical regression models for QoE estimation and prediction in audiovisual communications. IEEE Trans. Multimedia **19**(6), 1195–1208 (2017)
23. Torres Vega, M., De Vleeschauwer, D., Brunnström, K.: A review of machine learning applications in QoE management of multimedia services. IEEE Trans. Netw. Serv. Manage. **12**(2), 366–379 (2015)
24. Vasilev, V., et al.: Predicting QoE factors with machine learning. In: IEEE International Conference on Communications (ICC), pp. 1–6 (2018)
25. Zhou, W., et al.: A brief survey on adaptive video streaming quality assessment. J. Vis. Commun. Image Represent. **86**, 103526 (2022)

Performance Paradoxes in Matching Systems are not that Rare

A. Busic[1], J.M. Fourneau[1,2(✉)], A. Lunven[1], and S. Li[1]

[1] INRIA ARGO, Paris, France
{ana.busic,antoine.lunven,shu.li}@inria.fr
[2] DAVID, University Paris-Saclay, UVSQ, Versailles, France
Jean-Michel.Fourneau@uvsq.fr

Abstract. A Matching model describes the waiting times suffered by items before they match with other items and disappear immediately without service. The matching relation is described by a compatibility graph. It is an easy representation of multiple types synchronizations between items. When we add an edge in the compatibility graph, one expects that the expected total number of items decreases. Unexpectedly this is not always true and there is a performance paradox. Extending previous results, we show new matching models with performance paradoxes which seems to be more frequent than one can expect.

1 Introduction

Following [1], a Matching model is a triple (G, Φ, α) defined by

1. a matching graph $G = (\mathcal{V}, \mathcal{E})$, which is an undirected graph whose nodes in $\mathcal{V}$ are classes of items and whose edges in $\mathcal{E}$ model the allowed matching of items. G is called the compatibility graph or the Matching graph. Upon arrival, an item is queued if there are no compatible items present in the system.
2. Φ is a matching policy. It states the couple of items which is chosen upon arrival when an arriving item matches one or more items already waiting. Typical matching disciplines are First Come First Match (FCFM, an analog of First Come First Served in this approach) or Match the Longest Queue. Once they are matched, both items leave the system immediately (no need for service).
3. A description of the stochastic processes to model the arrivals. It can be a collection of independent Poisson processes for a continuous-time model or a distribution of probability over the types of items in discrete time. In this paper, we assume a continuous time model. The arrival of type i items follow a Poisson process with rate λ_i.

The general matching model proposed in [1] and [2] was considering a general undirected matching graph G and it is assumed that the arrivals of items occur one at a time. It is important to avoid the confusion with Bipartite Matching

Model (see for instance [3] and references therein) where the matching graph is bipartite and two items of distinct classes arrive at the same time.

General matching models are easy to describe but they not so simple to analyze. Assuming independent Poisson arrivals of items, and FCFM discipline the model is associated with an infinite Markov chain. A necessary condition of stability and a product form solution were proved in [2] and [1] for these assumptions.

Some of us have recently established that there exists some performance paradox for FCFM matching models [4]. When one adds new edges in the compatibility graph, one may expect that the expectation of the total number of items decreases. In [4], some examples were presented to show that it is not always the case and a sufficient condition for such a performance paradox to exist is proved. Thus, adding flexibility on the matching graph does not always result for performance improvement. The example was based on an almost complete graph. In [5], the results have been extended in many directions: all greedy disciplines instead of FCFM, and an infinite family of compatibility graphs. However some questions remain about the compatibility graphs which lead to performance paradoxes. Note that performance paradoxes have also been shown recently for bipartite matching models [6].

An application of Matching models studied in the literature was the kidney exchanges [7–9]. The kidney exchange arises when a healthy person who wishes to donate a kidney is not compatible (blood types or tissue types) with the receiver. Two incompatible pairs (or maybe more) can form a cyclic exchange, so that each patient can receive a kidney from a compatible donor.

The technical part of the paper is as follows. We begin in the next section with a short presentation of performance paradoxes for Matching systems. Section 3 is devoted to simple examples built for compatibility graphs with 5 and 7 nodes which were not considered previously. In Sect. 4, we also find some performance paradoxes for multi-graphs. The objective of the paper is to show arrival rates and compatibility graphs which exhibit performance paradoxes. The vector of arrival rates has been obtained through brute force inspection. Two numerical tools based on the closed form solution for expectation of total number of items based on the independent sets have been developed for cross-validation and a simulator was also used during the validation process [10,11]. In all experiments reported in Sect. 3 and 4, the arrival rates are not truncated or rounded up while the expectations are given with the precision of our Python codes.

2 Performance Paradoxes in Matching Systems

Let $G = (\mathcal{V}, \mathcal{F})$ be the matching graph. Usually, the nodes in $\mathcal{V}$ are also denoted as letters and an ordered list of letters is called a word in [1]. In this paper, we only consider FCFM matching disciplines even if some results on paradoxes have been established for the larger set of greedy disciplines [5]. To model this discipline, states are represented by ordered list of letters in their order of arrivals: a word. Assume that x is a letter from $\mathcal{V}$, we use the following notation:

- $\Gamma(x)$ is the set of neighbors of x in G.
- $\mathbb{E}[G,\lambda]$ will denote the expected number of items for compatibility graph G and vector of arrival rates λ.
- $|S|$ is the cardinal of set S. Let G be a graph and a,b two nodes of $\mathcal{V}$ such that (a,b) is not an edge of G, we will denote as $G+e$, the graph G where we have added edge $e=(a,b)$.
- Similarly, $G-(u,v)$ where (u,v) is an edge of G, is the graph obtained after removing edge (u,v).
- $\mathcal{I}$ will denote an independent set while the set of Independent Sets of the compatibility graph will be denoted as $\mathbb{I}$.
- Finally, $\alpha(S)$ is the sum of the arrival rates for letters in set S:

$$\alpha(S) = \sum_{i \in S} \lambda_i.$$

We begin by some results from the literature (see [1,2,4]).

Theorem 1. *[1] A necessary condition of stability under the FCFM discipline that $\alpha(\mathcal{I}) < \alpha(\Gamma(\mathcal{I}))$ for every independent set $\mathcal{I}$ in $\mathbb{I}$.*

Corollary 1. *Markov models associated with bipartite compatibility graphs cannot be ergodic.*

Theorem 2 (Product Form for the Steady-State Distribution [2]). *Assuming ergodicity, the steady state distribution for word w is:*

$$\pi(w) = C \frac{\prod_{i=1}^{l(w)} \alpha(\{w_i\})}{\prod_{k=1}^{l(w)} \alpha(\Gamma(\{w_1,..w_k\}))},$$

where $l(w)$ is the length of word w and w_i is the i-th letter of w. C is the normalizing constant and it is equal to $\pi(\emptyset)$, the probability of the empty word but it must be computed by normalization.

It remains to compute the normalizing constant (i.e. C) and the expectation of the total number of items in the system: a difficult task as the state space of the Markov chain is infinite. Closed-form expressions have been obtained for both quantities. We only report the result for the expectation.

Corollary 2 (Expectation [4]). *For all independent sets $\mathcal{I} \in \mathbb{I}$, consider an ordered version of $\mathcal{I}$ noted $\mathcal{I}^o = \{i_1, \cdots, i_{|\mathcal{I}|}\}$, we note σ a permutation of its elements, i.e. $\mathcal{I}^{\sigma(o)} = \{i_{\sigma(1)}, \cdots, i_{\sigma(|\mathcal{I}|)}\}$ and $\mathfrak{S}_{|\mathcal{I}|}$ the set of all permutations on the nodes of $\mathcal{I}$. We define:*

$$T_{\mathcal{I}^o} = \prod_{k=1}^{|\mathcal{I}|} \frac{\alpha(\{i_k\})}{\alpha(\Gamma(\{i_1,\cdots,i_k\})) - \alpha(\{i_1,\cdots,i_k\})}.$$

Furthermore $T_\mathcal{I} = \sum_{\sigma \in \mathfrak{S}_{|\mathcal{I}|}} T_{\mathcal{I}^{\sigma(o)}}$. We also define:

$$E_{\mathcal{I}^o} = \sum_{l=1}^{|\mathcal{I}|} \frac{\alpha(\Gamma(\{i_1, \cdots, i_l\}))}{\beta(\{i_1, \cdots, i_l\})} \prod_{k=1}^{|\mathcal{I}|} \frac{\alpha(\{i_k\})}{\beta(\{i_1, \cdots, i_k\})}$$

where $\beta(\{i_1, \cdots, i_l\}) = \alpha(\Gamma(\{i_1, \cdots, i_l\})) - \alpha(\{i_1, \cdots, i_l\})$.

Finally, $E_\mathcal{I} = \sum_{\sigma \in \mathfrak{S}_{|\mathcal{I}|}} E_{\mathcal{I}^{\sigma(o)}}$. $\mathbb{E}[G, \alpha]$ can be rewritten as a finite sum over all independent sets, i.e.

$$\mathbb{E}[G, \alpha] = \left(1 + \sum_{\mathcal{I} \in \mathbb{I}} T_\mathcal{I}\right)^{-1} \left(\sum_{\mathcal{I} \in \mathbb{I}} E_\mathcal{I}\right) \tag{1}$$

Computing the expectation is based on two summations: the inner one on the independent sets and the outer one on the permutation of the letters of the independent sets. Remark that as the number of independent sets can be exponential on $|\mathcal{V}|$, such an algorithm is in practice limited to compatibility graph with a small number of nodes.

Definition 1 (Performance Paradox). *For some arrival rates λ, we have a performance paradox, if $\mathbb{E}[G + e, \lambda] > \mathbb{E}[G, \lambda]$.*

Proposition 1 (First Performance Paradox [4]). *Let K_4 be the complete graph with 4 nodes. For some arrival rates λ, we have proved that*

$$\mathbb{E}[K_4 - (1,2), \lambda] < \mathbb{E}[K_4, \lambda].$$

The analysis in [4] requires one combinatorial notion. The bottleneck set, denoted $BS(G, \lambda)$ throughout this paper, is defined as

$$BS(G, \lambda) = \{\mathcal{I} \in \mathbb{I} \mid argmin(\alpha(\Gamma(I)) - \alpha(I))\}.$$

As it is difficult to use the formula on the expectation, we have developed in [5] some "lifting theorems" to explain how one can build large compatibility graphs and arrival rates based on small compatibility graphs. We have proved that the existence of performance paradoxes on small graphs implies the existence of such paradoxes on large graphs. Then from the initial seed graph (i.e. $K_4 - (1,2)$ in [4]) we have obtained an infinite set of graphs with paradoxes with the help of the JOIN operation defined as follows.

Definition 2 (JOIN $\bowtie$). *Let $G1 = (V1, E1)$ and $G2 = (V2, E2)$ two arbitrary graphs, the join of $G1$ and $G2$ (denoted as $G1 \bowtie G2$) is defined by the set of nodes $V1 \cup V2$ and the set of edges $E1 \cup E2 \cup V1 \times V2$, where $\times$ is the Cartesian product.*

Proposition 2 (Construction of Performance Paradox [5]). *Performance paradoxes exist for graphs $IN_n \bowtie (K_4 - (1,2))$ where IN_n is the graph with n isolated nodes.*

From these results, some interesting questions arise:

Remark 1.

- Note that all nodes in IN_n in the previous construction are stars in the compatibility graph (i.e. they are connected to all nodes). This property results from the definition of the JOIN operation. Thus one may ask if a star in the compatibility graph is needed to obtain a performance paradox.
- The graphs exhibiting paradoxes in Property 2 are not dense because the average degree may be as small as 4. But one may ask if it is possible to find paradoxes with compatibility graph the average degree of which is almost 1.

We address both questions in the next sections and we show that a star is not needed and the compatibility graph may have a degree equal to 1.

3 Star-Free Graphs

We say that a graph G is *star-free* if G contains no induced subgraph isomorphic to a star. Remember that bipartite compatibility graphs do not lead to ergodic Markov chains. Therefore, one must chose a small compatibility graph G to have a small number of Independent Sets but such that G is not bipartite. Furthermore G must be star-free and connected. Note that both properties are monotone. Thus if these properties hold for G, they also hold for $G+e$. Furthermore G must not be complete. An enumeration argument shows that the smallest star-free, non-bipartite graphs contain one odd cycle and five nodes. Thus, we begin with the following two graphs on 5 nodes: cycle C_5 and F_1 as depicted in Fig. 1. Next, we enlarge the vertex set by one and study F_3, obtained from C_6 by inserting an edge. It is the smallest example in which two distinct odd cycles coexist. Finally, we study the 7-cycle C_7 with 28 independent sets. The arrival rates leading to paradoxes are found by exhaustive search on the sets of parameters which insure the stability.

3.1 Graphs C_5, F_1 and F_2

Clearly, graph C_5 has 5 nodes and 5 edges, it is connected, and the number of independent sets is 10. Its average degree is 1. It is possible to use the closed form solution to compute $\mathbb{E}[C_5, \lambda]$ for any vector of arrival rates. Graph F_1 is shown in Fig. 1. By inserting the edges $\{x_2, x_5\}$ in C_5 and $\{z_4, z_5\}$ in F_1, respectively, we obtain a graph isomorphic to the house graph, denoted F_2. In what follows we analyze cycle graph C_5, graph F_1, and house graph F_2.

Using the Python codes in [10], one obtains the following results. For cycle C_5, and arrival rate vector $(0.111, 0.10575, 0.179426, 0.35046, 0.253364)$, replacing it by F_2 raises the expectation from 7.86 to 8.17. A similar upward shift is observed when graph F_1 with arrival rate vector $(0.266025, 0.11005, 0.210855, 0.14145, 0.27162)$ is augmented to F_2. The expected total number of items grows from 16.069 to 16.89. The Bottleneck independent set is $\{v_1, v_4\}$ for the first example and $\{v_3, v_5\}$ for the second one (Fig. 2).

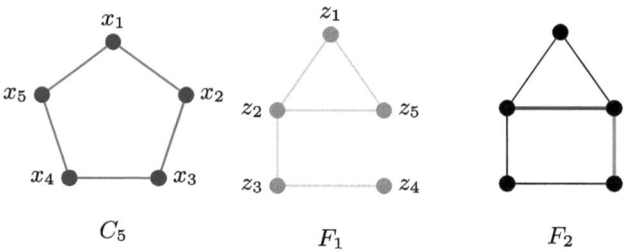

Fig. 1. Graphs: C_5, F_1 and F_2.

3.2 F_3: A Star-Free Graph with Two Odd Cycles

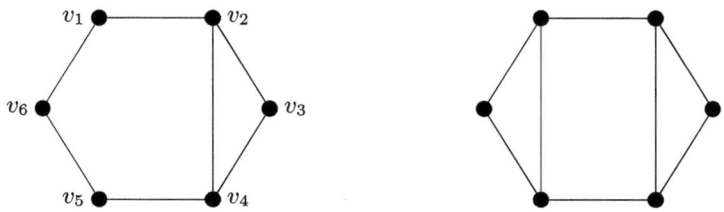

Fig. 2. F_3 and $F_3 + \{v_1, v_5\}$.

As noted previously, graph F_3 is obtained by inserting edge $\{v_2, v_4\}$ into graph C_6. The resulting star-free graph contains two distinct odd cycles with lengths 3 and 5, with 15 independent sets. With rates $\lambda_1 = 0.231787$, $\lambda_2 = 0.462366$, $\lambda_3 = 0.14472$, $\lambda_4 = 0.127484$, $\lambda_5 = 0.00342$ and $\lambda_6 = 0.030223$, we get with the Python codes described in [10] and [11] an expectation of total number of items equal to 31.59 for graph F_3. This expectation increases to 31.71 when we add edge $\{v_1, v_5\}$, again a performance paradox. Both graphs have bottleneck independent set equal to $\{v_2, v_6\}$.

3.3 C_7: Performance Paradox on a Simple Cycle

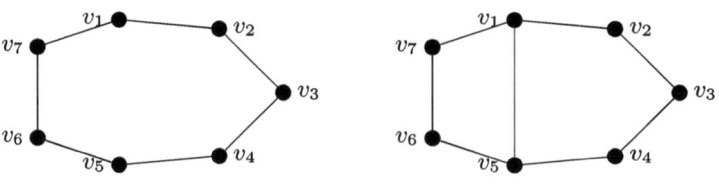

Fig. 3. C_7 and $C_7 + \{v_1, v_5\}$.

Graph C_7 is a simple odd cycle, with 28 independent sets. We study cycle graph C_7 together with graph $C_7 + \{v_1, v_5\}$, as illustrated in Fig. 3. Using the Python code in [10], we select the arrival rate vector

$$\lambda = [0.045206, 0.240384, 0.303789, 0.146114, 0.132526, 0.111772, 0.020209].$$

For this choice of parameters we obtain $\mathbb{E}[C_7, \lambda] = 129.56$ and $\mathbb{E}[C_7 + \{v_1, v_5\}, \lambda] = 137.87$. Both graphs share the same bottleneck independent set $\{v_2, v_4.v_6\}$.

4 (Multi)-graph with Self Loops

Here, we assume that some nodes in the matching graph have a self-loop and this was clearly forbidden by the previous assumptions in [1]. Thus the compatibility graph becomes a multigraph. Note that the results obtained in [2] and [1] are not valid anymore because of some technical details in the proof of product form. Such a model may represent how we can organize fair competition between players with roughly the same ranking (for instance ELO points for chess). The theory for matching models with multigraph has recently been presented in [12] where the authors proved a closed form solution for the steady-state distribution when it exists. A less general result has been presented independently in [13] where all the nodes of the compatibility graph have a self loop. In that last case, the Markov chain is finite and the existence of the stationary distribution is not a problem anymore.

4.1 Product Form on Multigraphs

Let $G = (\mathcal{V}, \mathcal{E})$ be a multigraph with the nodes partition $\mathcal{V} = \mathcal{V}_1 \cup \mathcal{V}_2$ where $\mathcal{V}_1 = \{i \in \mathcal{V} \mid \{i,i\} \in \mathcal{E}\}$ and $\mathcal{V}_2 = \mathcal{V} \setminus \mathcal{V}_1$. Denote by $\mathbb{I}(\check{G})$ the family of independent sets of $\check{G} = (\mathcal{V}, \mathcal{E} \setminus \{(i,i) \mid i \in \mathcal{V}_1\})$.

Theorem 3 (Product form [12, Theorem 4.5]). *Assume FCFM policy and ergodicity. For any word $w = (w_1, \ldots, w_q) \in W$,*

$$\pi(w) = \pi(\emptyset)\bar{\pi}(w) = \pi(\emptyset) \prod_{l=1}^{q} \frac{\alpha(\{w_l\})}{\alpha\big(\Gamma(\{w_1, \ldots, w_l\})\big)}, \qquad (2)$$

where

$$\pi(\emptyset)^{-1} = 1 + \sum_{\mathcal{I} \in \mathbb{I}(\check{G})} \sum_{\sigma \in \mathfrak{S}_{|\mathcal{I}|}} \prod_{k=1}^{|\mathcal{I}|} \frac{\alpha(\{e_{\sigma(k)}\})}{\alpha\big(\Gamma(\{e_{\sigma(1)}, \ldots, e_{\sigma(k)}\})\big) - \alpha(\{e_{\sigma(1)}, \ldots, e_{\sigma(k)}\} \cap \mathcal{V}_2)}.$$

Ref [12] does not provide a closed form solution for the expectation. Thus we have developed such a formula based on the ideas proposed in previous works [2] and [4]. Clearly,

$$\mathbb{E}[G, \alpha] = \sum_{w \in W} |w|\, \pi(w) = \pi_0 \sum_{w \in W} |w|\bar{\pi}(w) = \pi_0 \sum_{w \neq \emptyset} |w|\bar{\pi}(w).$$

The summation is on the set of states which is infinite but highly structured. We now rewrite the sum over all words w as a finite sum over independent sets of $\check{G}$. For $\mathcal{I} \in \mathbb{I}(\check{G})$, fix an arbitrary ordering $\mathcal{I}^o = \{e_1, \ldots, e_{|\mathcal{I}|}\}$. Given a permutation σ of the index set $\{1, \ldots, |\mathcal{I}|\}$, define the reordered list $\mathcal{I}^{\sigma(o)} = \{e_{\sigma(1)}, \ldots, e_{\sigma(|\mathcal{I}|)}\}$, that is, the sequence obtained by applying σ to the ordered version $\mathcal{I}^o$. We denote $\mathcal{I}_k^o$ as the first k elements of $\mathcal{I}^o$. Let $\mathcal{J}_k^o = \mathcal{I}_k^o \cap \mathcal{V}_2$. We define

$$T_{\mathcal{I}^o} = \prod_{k=1}^{|\mathcal{I}|} \frac{\alpha(\{e_k\})}{\alpha\big(\Gamma(\mathcal{I}_k^o)\big) - \alpha\big(\mathcal{J}_k^o\big)},$$

and $T_{\mathcal{I}} = \sum_{\sigma \in \mathfrak{S}_{|\mathcal{I}|}} T_{\mathcal{I}^{\sigma(o)}}$. Thus $\pi(\emptyset) = (1 + T_{\mathcal{I}})^{-1}$. We also define

$$E_{\mathcal{I}^o} = \sum_{l=1}^{|\mathcal{I}|} \frac{\alpha\big(\Gamma(\mathcal{I}_l^o)\big)}{\alpha\big(\Gamma(\mathcal{I}_l^o)\big) - \alpha\big(\mathcal{J}_l^o\big)} \prod_{k=1}^{|\mathcal{I}|} \frac{\alpha(\{e_k\})}{\alpha\big(\Gamma(\mathcal{I}_k^o)\big) - \alpha\big(\mathcal{J}_k^o\big)},$$

and $E_{\mathcal{I}} = \sum_{\sigma \in \mathfrak{S}_{|\mathcal{I}|}} E_{\mathcal{I}^{\sigma(o)}}$.

Corollary 3 (Expectation). *The expected number of items in steady state is*

$$\mathbb{E}[G, \alpha] = \Big(1 + \sum_{\mathcal{I} \in \mathbb{I}(\check{G})} T_{\mathcal{I}}\Big)^{-1} \sum_{\mathcal{I} \in \mathbb{I}(\check{G})} E_{\mathcal{I}}. \tag{3}$$

Proof. The proof is omitted for the sake of readability and will be presented in an extended version of this paper which will be available on the web.

Remark 1. *Setting $\mathcal{V}_2 = \emptyset$, Theorem 2 together with Corollary 2 are recovered as special instances of Theorem 3 and Corollary 3, respectively.*

4.2 Examples of Graphs with 4 or 7 Nodes

Fig. 4. Graphs: G and G_2.

We consider $G = (\mathcal{V}, \mathcal{E})$ in Fig. 4, where $\mathcal{V}$ is partitioned into $\mathcal{V}_1 = \{v_2, v_3, v_4\}$ and $\mathcal{V}_2 = \{v_1\}$. The independent sets of $\check{G}$ are $\{\{v_1\}, \{v_2\}, \{v_3\}, \{v_4\}, \{v_2, v_4\}, \{v_2, v_3\}\}$. Graph G_2 is obtained by inserting the

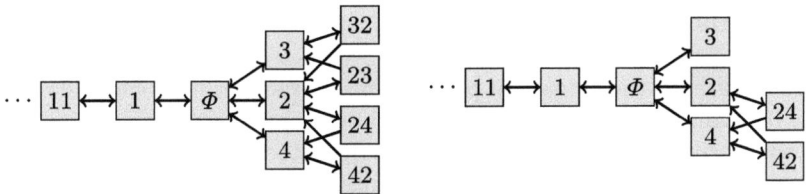

Fig. 5. Truncation of the Markov Chains for G (left) and G_2 (right).

edge $\{v_2, v_3\}$ (see Fig. 4). The associated Markov Chains, truncated to states with at most two letters, are depicted in Fig. 5. The rates are $\lambda_1 = 0.432685$, $\lambda_2 = 0.005861$, $\lambda_3 = 0.141128$ and $\lambda_4 = 0.420326$. The BS set is a singleton (i.e. $\{v1\}$). We have a paradox as the expectation slightly increases from 2.944793 to 2.947256 when we add the edge.

The former example does not address the case of adding a new edge e connecting a node lacking a self-loop to a node with self-loop. To fill this gap, we study graph H_2 (see Fig. 6), obtained from H by inserting edge $\{v_1, v_5\}$. Similarly, graph H_1 is formed by connecting v_2 and v_3, while graph H_3 is obtained by adding a self-loop to v_5. We have found performance paradoxes in all cases. For arrival rate vector $(0.070716, 0.180906, 0.238414, 0.204521, 0.164007, 0.07243, 0.069006)$ the bottleneck set is $\{v_7\}$. The expectation is 31.218542 for graph H and it increases to 31.539906 for graph H_1. Similarly, if we start with arrival rate vector $(0.08502, 0.06418, 0.076071, 0.096493, 0.336406, 0.326572, 0.015258)$ for H, we get an expectation for the total number equal to 13.0179 and the expectation is equal to 13.058815 for the same rates and graph H_2. Finally, the comparison between H and H_3 for rate vector $(0.104627, 0.052725, 0.194586, 0.190934, 0.117413, 0.225466, 0.114249)$ also shows a performance paradox: the expectation was 33.838547 for H and it jumps to 35.972599 for H_3. In both cases, the bottleneck independent set is singleton $\{v_6\}$.

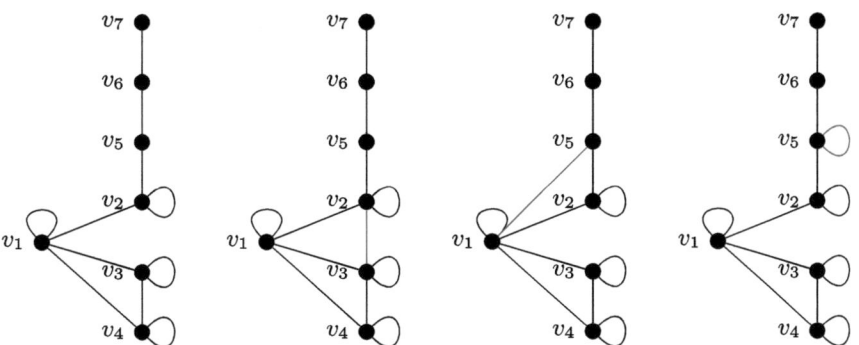

Fig. 6. graphs: H, H_1, H_2 and H_3.

5 Conclusions and Remarks

This study extends the catalog of performance paradoxes in stochastic matching models by exhibiting new star-free, sparsely connected graphs (some with self-loops) where adding a single compatibility edge or self loop increases the steady-state number of waiting items. These results demonstrate that the paradox is structurally robust, arising even in small odd-cycle graphs such as F_3 and C_7. Beyond enriching the theoretical landscape, the findings caution practitioners that naive flexibility enhancements may backfire in real-world platforms.

References

1. Mairesse, J., Moyal, P.: Stability of the stochastic matching model. J. Appl. Probab. **53**(4), 1064–1077 (2016)
2. Moyal, P., Bušić, A., Mairesse, J.: A product form for the general stochastic matching model. J. Appl. Probab. **58**(2), 449–468 (2021)
3. Bušić, A., Gupta, V., Mairesse, J.: Stability of the bipartite matching model. Adv. Appl. Probab. **45**(2), 351–378 (2013)
4. Cadas, A., Doncel, J., Fourneau, J.-M., Bušić, A.: Flexibility can hurt dynamic matching system performance. SIGMETRICS Perform. Eval. Rev. **49**(3), 37–42 (2022)
5. Bušić, A., Cadas, A., Doncel, J., Fourneau, J.-M.: Performance paradox of dynamic matching models under greedy policies. Queu. Syst. **107**(3–4), 257–293 (2024)
6. Iriondo, I., Doncel, J.: Performance paradox of dynamic bipartite matching models. In: *Network Games, Artificial Intelligence, Control and Optimization - 11th International Conference, NETGCOOP 2024, France*, vol. 15185 of *Lecture Notes in Computer Science*, pp. 47–56. Springer (2024)
7. Utku Ünver, M.: Dynamic kidney exchange. Rev. Econ. Stud. **77**(1), 372–414 (2010)
8. United Network for Organ Sharing. Living donation kidney paired. https://unos.org/wp-content/uploads/unos/livingdonationkidneypaired.pdf
9. Ashlagi, I., Jaillet, P., Manshadi, V.H.: Kidney exchange in dynamic sparse heterogenous pools. In: *Proceedings of the Fourteenth ACM Conference on Electronic Commerce*, EC '13, pp. 25–26 (2013)
10. Li, S.: Matching paradox solver (2025). https://github.com/shuligraph/matching-paradox-sim
11. Lunven, A.: Matching paradox simulator (2025). https://github.com/Antoine-lunven/Performance-paradoxes-in-Matching-systems-are-not-that-rare/tree/main
12. Begeot, J., Marcovici, I., Moyal, P., Rahme, Y.: A general stochastic matching model on multigraphs. Latin Am. J. Probab. Math. Stat. **18**(1), 1325 (2021)
13. Busic, A., Cadas, A., Doncel, J., Fourneau, J.M.: Product form solution for the steady-state distribution of a Markov chain associated with a general matching model with self-loops. In: *Computer Performance Engineering - 18th European Workshop, EPEW 2022, Spain*, vol. 13659 *Lecture Notes in Computer Science*, pp. 71–85. Springer (2022)

An Anti-eavesdropping Strategy in Communication with a Group of Nodes

Andrey Garnaev[✉] and Wade Trappe

WINLAB, Rutgers University, North Brunswick, NJ, USA
garnaev@yahoo.com, trappe@winlab.rutgers.edu

Abstract. The wireless communication between a transmitter and a group of nodes, such as a control center interacting with several drones engaged in a specific mission, may be vulnerable to malicious eavesdropping attacks. Cons of traditional communication protocol aimed to maximize total secrecy rate is that might lead to neglecting communication with some nodes, and, to failure of nodes' mission. To deal with such cons we suggest a maximizing minimal secrecy rate protocol, or, in other words, a guaranteed (expected) secrecy rate protocol. The problem is modeled and solved in a (Bayesian) game-theoretical framework between a transmitter and an adversary. We prove that such protocol supports uninterrupted communication and maintains its stability even if the transmitter has incomplete information on what type of antenna (directional or omni-directional) the adversary has. Finally, an approach to solve an adversary dilemma of which of the antennas to engage is suggested and illustrated.

Keywords: Eavesdropping · Max-Min · Equilibrium

1 Introduction

Ensuring secrecy and confidentiality in communication between a transmitter and receivers is a critical challenge in the development of secure communication systems. Eavesdropping represents one method through which this confidentiality can be compromised. The problem of secure communication in such scenarios involves different agents (e.g., a transmitter and an adversary). An adversary might be: (I) a *passive* agent, passively listening in ongoing transmission, or (II) an *active* agent. In the last case, game theory has been employed to model and analyze such problems [11]. The adversary might be active by combining passive eavesdropping to eavesdrop signal and active jamming to degrade the signal at the receiver [1,5,7,10,16]. The adversary can also be active by choosing a position for the adversary to eavesdrop upon the ongoing communication [3].

Using directional antennas in a multi-receiver environment might create an extra possibility for both agents to be active: (A) to improve secrecy in communication for the transmitter [4]; (B) to improve eavesdropping capacity for the adversary [14]. In [9], an anti-eavesdropping strategy was derived, where the transmitter not only seeks to maximize the secrecy communication rate but also aims to do so in a manner that is least predictable to the eavesdropper via choosing with which receiver to communicate.

In this paper, we consider a multi-receiver system in the presence of an eavesdropper, modeling a control center that must establish communication with n drones involved in carrying out a mission. Motivated by the observation that (traditional) communication protocols aimed to maximize total secrecy rate might lead to neglecting communication with some nodes, and, to the failure of nodes' mission, in this paper, we suggest a maximizing the minimal secrecy rate, or, in other words, a guaranteed (expected) secrecy rate protocol. The problem is modeled and solved in a (Bayesian) game-theoretical framework between a transmitter and an adversary. Equilibrium strategies are derived, and their uniqueness is proven. The advantages of the introduced protocol in comparison with the maximizing total secrecy rate protocol are illustrated.

2 Communication Model

In this paper, we consider a *transmitter* which must communicate with n nodes secretly in the presence of an adversary. The transmitter employs a directional antenna for communication with each of the n nodes. For instance, one might consider a control center that must establish communication with n drones involved in carrying out a mission. It is worth noting that an antenna is a device that is used for radiating/collecting radio signals into/from space. Say, an omni-directional antenna, which can radiate/collect radio signals uniformly to all directions in space. Different from an omni-directional antenna, a directional antenna can concentrate transmitting or receiving capability to some desired directions so that it has better performance than an omni-directional antenna. Thus, a pros of a directional antenna scheme is that it could allow the detection of weaker or more distant signals and might reduce interference from other sites [13]. In the scenario considered in this paper, an *adversary* intends to eavesdrop upon communication of the transmitter with group of the nodes, and it might be equipped either with directional antenna or omni-directional antenna to eavesdrop upon the nodes. It is worth noting that besides possible improvement of eavesdropping capacity, the directional antenna turns the adversary from a passive agent to an active agent, as successful eavesdropping demands the selection of a specific node to focus on.

(A) **Let the adversary be an active agent**, i.e., it is equipped with a directional antenna. Then:
(A-I) If both transmitter and adversary select the same node, say, node i, to communicate and eavesdrop upon, the secrecy rate (capacity) [8,16] of such communication can be given as follows:

$$C_{S,i} \triangleq \lfloor \ln(1 + h_i P) - \ln(1 + h_{E,i} P) \rfloor_+ \quad (1)$$

with $\lfloor \xi \rfloor_+ \triangleq \max\{\xi, 0\}$, h_i is the channel gain between the transmitter and node i, $h_{E,i}$ is the eavesdropping channel gain between node i and the adversary, and P is the power level applied by the transmitter;

(A-II) If the transmitter and adversary select different nodes, say, node i and node j, respectively, with $i \neq j$ then the capacity of transmitter-node i channel in such absence of eavesdropping attack on node i, is

$$T_i \triangleq \ln(1 + h_i P). \tag{2}$$

(B) **Let the adversary be a passive agent**, i.e., it is equipped, say, with an omni-directional antenna. Then if transmitter selects node i to communicate with, the secrecy rate (capacity) of such communication can be given as follows:

$$C_{S,i}^0 \triangleq \lfloor \ln(1 + h_i P) - \ln(1 + h_{E,i}^0 P) \rfloor_+ \tag{3}$$

with $h_{E,i}^0$ is the eavesdropping channel gain between node i and adversary, and

$$h_{E,i} > h_{E,i}^0 \text{ for } i \in \mathcal{N}, \tag{4}$$

what reflects worse eavesdropping capability of omni-directional antenna in comparison with directional antenna.

Note that to achieve secure communication, the transmitter must disregard any channels that do not serve this purpose. This excludes any channel i with zero secrecy rate. That is why, by (1) and (4), without loss of generality, we assume that the following relation holds:

$$h_i > h_{E,i} \text{ for } i \in \mathcal{N}. \tag{5}$$

A transmitter's strategy is a vector of probabilities $\boldsymbol{x} = (x_1, \ldots, x_n)$, where x_i is the probability of selecting node i to communicate with. Thus, the set of transmitter's feasible strategies is the set of n-dimension probability vectors and we denote this set by $\mathcal{P}$, i.e., $\boldsymbol{x} \in \mathcal{P}$.

Payoff to the Transmitter: In this paper, we consider the most dangerous scenario for the transmitter in which it does not know which type of antenna the adversary has, or in other words, whether the adversary is an active agent or a passive agent. Specifically, the transmitter only knows that:

(P-I) *the adversary is an active agent* with probability γ. In this case its strategy is a probability vector $\boldsymbol{y} = (y_1, \ldots, y_n)$, where y_i is the probability of selecting node i to eavesdrop upon. Thus, the set of (active) adversary's feasible strategies also is the set of n-dimension probability vectors $\mathcal{P}$, i.e., $\boldsymbol{y} \in \mathcal{P}$.

(P-II) *the adversary is a passive agent* with probability $1 - \gamma$.

By (P-I), (P-II) and (1)–(5), we have that, if the transmitter selects node i to communicate with and (active) adversary applies strategy $\boldsymbol{y} = (y_1, \ldots, y_n)$, then the expected secrecy rate $\mathcal{C}_{S,i}(\boldsymbol{y})$ is

$$\mathcal{C}_{S,i}(\boldsymbol{y}) \triangleq \gamma(C_{S,i}y_i + T_i(1 - y_i)) + (1 - \gamma)C_{S,i}^0$$
$$= T_i - \gamma E_i y_i - (1 - \gamma)E_i^0, \qquad (6)$$

where E_i and E_i^0 are "eavesdropped" rates at node i for an active and passive adversary, respectively, given as follows:

$$E_i \triangleq T_i - C_{S,i} \text{ and } E_i^0 \triangleq T_i - C_{S,i}^0, \qquad (7)$$

By (4) and (5), "eavesdropped" rates E_i and E_i^0 at node i are positive.

By (6), we have that if transmitter applies strategy $\boldsymbol{x} = (x_1, \ldots, x_n)$ and (active) adversary applies strategy $\boldsymbol{y} = (y_1, \ldots, y_n)$, then

(A) the (expected) *minimal* secrecy rate of the transmitter is

$$\mathcal{C}_S^M(\boldsymbol{x}, \boldsymbol{y}) \triangleq \min_{i \in \mathcal{N}} x_i \mathcal{C}_{S,i}(\boldsymbol{y}) = \min_{i \in \mathcal{N}} x_i(T_i - \gamma E_i y_i - (1 - \gamma)E_i^0); \qquad (8)$$

(B) the (expected) *total* secrecy rate of the transmitter is

$$\mathcal{C}_S^T(\boldsymbol{x}, \boldsymbol{y}) \triangleq \sum_{i \in \mathcal{N}} x_i \mathcal{C}_{S,i}(\boldsymbol{y}) = \sum_{i \in \mathcal{N}} x_i(T_i - \gamma E_i y_i - (1 - \gamma)E_i^0). \qquad (9)$$

Since the total secrecy rate (9) is linear on $\boldsymbol{x}$, it leads to a strategy maximizing it focused only on the nodes in which $T_i - \gamma E_i y_i - (1 - \gamma)E_i^0$ achieves its maximum over $i \in \mathcal{N}$. Such an optimizing strategy might not maintain reliable communication with each node engaged in the mission. Thus, it might lead to a drastic consequence since communication with some nodes will be neglected (please see also Sect. 4 below) and the mission might fail as a whole.

To avoid such cons of maximizing total secrecy rate strategy, in this paper, we suggest considering the (expected) minimal secrecy rate of the transmitter $\mathcal{C}_S^M(\boldsymbol{x}, \boldsymbol{y})$, or in another word, the (expected) guaranteed secrecy rate for each node as the payoff to the transmitter.

Payoff to the Adversary: First note that, if the transmitter and (active) adversary apply strategies $\boldsymbol{x} = (x_1, \ldots, x_n)$ and $\boldsymbol{y} = (y_1, \ldots, y_n)$, respectively, then, by (7), the minimal "eavesdropped" rate is given as follows:

$$\mathcal{E}_A^M(\boldsymbol{x}, \boldsymbol{y}) \triangleq \min\{E_i x_i y_i : i \in \mathcal{N}\}. \qquad (10)$$

We consider this minimal "eavesdropped" rate $\mathcal{E}_A^M(\boldsymbol{x}, \boldsymbol{y})$ as the payoff to the (active) adversary. Meanwhile, $\min\{E_i^0 x_i : i \in \mathcal{N}\}$ is the payoff to the (passive) adversary.

Nash-Bayesian Equilibrium: The transmitter and adversary can be considered as *players* and we look for *(Nash-Bayesian) equilibrium* [6]. In other words, we look for such strategies x and y which are the best response to each other, i.e., they are a solution of the best response equations given as follows:

$$x = \text{BR}_T^M(y) \triangleq \text{argmax}\{\mathcal{C}_S^M(\tilde{x}, y) : \tilde{x} \in \mathcal{P}\}, \tag{11}$$

$$y = \text{BR}_A^M(x) \triangleq \text{argmax}\{\mathcal{E}_A^M(x, \tilde{y}) : \tilde{y} \in \mathcal{P}\}. \tag{12}$$

We call this (Bayesian) game the MM-game and denote it by Γ^M.

3 Equilibrium in Closed Form

In the following theorem using a constructive approach we derive equilibrium strategies in closed form and prove their uniqueness.

Theorem 1. *In the MM-game Γ^M there is a unique equilibrium. Moreover, the unique equilibrium strategies $x = (x_1, \ldots, x_n)$ and $y = (y_1, \ldots, y_n)$ of the transmitter, and adversary, respectively, are given as follows:*

$$x_i = \frac{1}{(T_i - (1-\gamma)E_i^0)\overline{X}} \text{ and } y_i = \frac{T_i - (1-\gamma)E_i^0}{E_i \overline{Y}} \text{ for } i \in \mathcal{N} \tag{13}$$

with

$$\overline{X} \triangleq \sum_{i \in \mathcal{N}} \frac{1}{T_i - (1-\gamma)E_i^0} \text{ and } \overline{Y} \triangleq \sum_{i \in \mathcal{N}} \frac{T_i - (1-\gamma)E_i^0}{E_i}. \tag{14}$$

Moreover, the payoffs for these equilibrium strategies are

$$\mathcal{C}_S^M(x, y) = \overline{XY}/(\overline{Y} - \gamma) \text{ and } \mathcal{E}_A^M(x, y) = \overline{XY}. \tag{15}$$

The proof of Theorem 1 can be found in the appendix.

A distinguishing feature of the derived transmitter's equilibrium strategy in Theorem 1 is that it is an egalitarian strategy in the sense that it maintains the same secrecy level for each node.

4 Total Secrecy and Eavesdropped Rates as Payoffs

In a traditional scenario, total secrecy and total "eavesdropped" rates are considered as payoffs. Specifically, the payoff to the transmitter is total secrecy rate $\mathcal{C}_S^T(x, y)$ given by (9), and the payoff to the (active) adversary is total "eavesdropped" rate

$$\mathcal{E}_A^T(x, y) \triangleq \sum_{i \in \mathcal{N}} E_i x_i y_i. \tag{16}$$

Meanwhile, $\sum_{i \in \mathcal{N}} E_i^0 x_i$ is the payoff to the (passive) adversary.

We call this (Bayesian) game the MT-game and denote it by Γ^T. Following [9] we can find its equilibrium $(\boldsymbol{x}, \boldsymbol{y}) = ((x_1, \ldots, x_n), (y_1, \ldots, y_n))$ as follows:

$$(x_i, y_i) = \begin{cases} \left(\dfrac{1/E_i}{\sum_{j \in I_+(\omega)} (1/E_j)}, \dfrac{T_i - (1-\gamma)E_i^0 - \omega}{E_i \gamma} \right) & i \in I_+(\omega), \\ (0, 0), & i \notin I_+(\omega) \end{cases} \quad (17)$$

with $I_+(\omega) \triangleq \{i \in \mathcal{N} : T_i > (1-\gamma)E_i^0 + \omega\}$ and ω is the unique positive root of the equation

$$\sum_{i \in \mathcal{N}} \frac{\lfloor T_i - (1-\gamma)E_i^0 - \omega \rfloor_+}{E_i} = \gamma. \quad (18)$$

5 Numerical Illustration and Discussion

Recall that the channel gains of the transmitter and (active and passive) adversary depend on their distances to the node via the path-loss factor $h = \tilde{h}/d^r$, $h_E = \tilde{h}_E/d_E^r$, $h_E^0 = \tilde{h}_E^0/d_E^r$, where d and d_E are the distances from the transmitter and the adversary to the node, $\tilde{h}$, $\tilde{h}_E$ and $\tilde{h}_E^0$ are corresponding fading coefficients for the transmitter, active adversary and passive adversary, respectively, and $r > 0$ is the path-loss factor. By this, we illustrate equilibrium strategies derived in Theorem 1 via an example of a network of $n = 4$ nodes, and the distances to the nodes allocated according to exponential law $d_i = \exp(ai)$ and

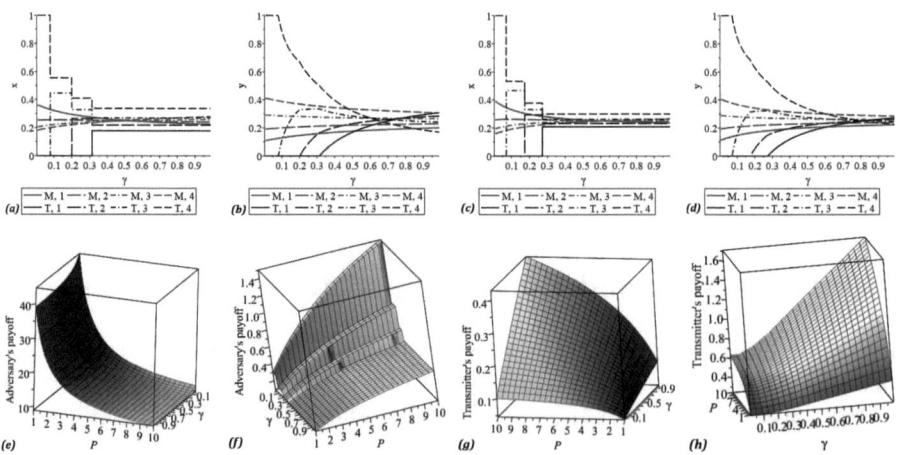

Fig. 1. (a) Transmitter's strategies for $P = 1$, (b) adversary's strategy for $P = 1$, (c) transmitter's strategies for $P = 10$, (d) adversary's strategy for $P = 10$, (e) adversary's payoff in game Γ^M, (f) adversary's payoff in game Γ^T, (g) transmitter's payoff in game Γ^M and (h) adversary's payoff in game Γ^T.

$d_{E,i} = \exp(a_E i)$, $i \in \mathcal{N}$ where $a = 0.095$, $a_E = 0.587$, $\tilde{h} = 1$, $\tilde{h}_E = 1.3$, $\tilde{h}_E^0 = 0.9$ and $r = 1$. Also, let applied transmission power P varies from 1 to 10.

Figure 1(a) and Fig. 1(c) illustrate that for MT-game Γ^T (please see Sect. 4) some nodes might be not communicated by the transmitter under some network's parameters. Specifically, node 1, node 2, and node 3 are not communicated if $\gamma < 0.8$ for $P = 1$ and if $\gamma < 0.7$ for $P = 10$ (please see also (17)). Meanwhile, in MM-game Γ^M, communication with all the nodes is maintained under any network's parameters (Theorem 1). This is reflected by the fact that none of the strategies becomes equal to zero (see also (13)). The other essential advantage of anti-eavesdropping strategy of MM-game Γ^M is that it is less sensitive to varying network parameters in comparison to anti-eavesdropping strategy of MT-game Γ^T. This is reflected by the observation that anti-eavesdropping strategy in MM-game is continuous on the network's parameters (please see Fig. 1(a), Fig. 1(c) and (13)), meanwhile, the anti-eavesdropping strategy in MT-game might have a jump on network's parameters (see (17), Fig. 1(a) and Fig. 1(c)). Adversary's strategies in both games are continuous on the network parameters (see Fig. 1(b), Fig. 1(d), (13) and (17)). Thus, in both games, adversary's strategies are flexible to varying network parameters. The other crucial advantage of the suggested anti-eavesdropping strategy in MM-game is that it is the most simple and convenient in implementation since it is given in closed form (see (13)) meanwhile, MT-game assumes solving a water-filling type equation (18) (about water-filling equations please see [2,15]). Figure 1(e)-(h) illustrates dependence of transmitter's payoffs in both games on probability γ and transmission power P. Since equilibrium strategies of both players in MM-game Γ^M are continuous on the network parameters, their payoffs are also continuous (see also (15)). Meanwhile, in MT-game Γ^T, adversary's payoff might have a jump in contrast to transmitter's payoff which is continuous on the network parameters. This is caused by the fact that adversary's payoff $\mathcal{E}_A^T(\boldsymbol{x}, \boldsymbol{y})$ is the sum of the product of continuous and discontinuous functions y_i and x_i, respectively, (please, see (17)).

How to Optimize Antenna's Selection: In the case of a repeating scenario, probability γ can be considered as a frequency to engage by the adversary one of two antenna types: either the directional or omni-directional antenna. So, γ can be considered as a control parameter for the adversary. One might wonder: *What can be the optimal frequency of such engagement for the adversary?* To response on this question let us introduce the utility reflecting total (integrated) payoff to the adversary for engaging both antennas:

(A) for MM-game Γ^M with $(\boldsymbol{x}_\gamma^M, \boldsymbol{y}_\gamma^M)$ as its equilibrium:

$$\mathcal{E}_{A,Total}^M(\boldsymbol{x}_\gamma^M, \boldsymbol{y}_\gamma^M) \triangleq \min_{i \in \mathcal{N}} x_{i,\gamma}^M (\gamma E_i y_{i,\gamma}^M + (1-\gamma) E_i^0); \tag{19}$$

(B) for MT-game Γ^T with $(\boldsymbol{x}_\gamma^T, \boldsymbol{y}_\gamma^T)$ as its equilibrium:

$$\mathcal{E}_{A,Total}^T(\boldsymbol{x}_\gamma^T, \boldsymbol{y}_\gamma^T) \triangleq \sum_{i \in \mathcal{N}} x_{i,\gamma}^T (\gamma E_i y_{i,\gamma}^T + (1-\gamma) E_i^0). \tag{20}$$

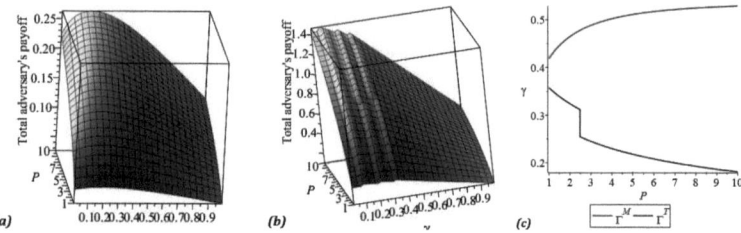

Fig. 2. (a) Total (integrated) adversary's payoff in game Γ^M, (b) total (integrated) adversary's payoff in game Γ^T, (c) trade-off γ.

Then, such γ, in which the total (integrated) payoff achieves its maximum, can be considered as the trade-off frequency of engaging both antennas by the adversary (Fig. 2(a)-(c)). This trade-off frequency can be found using the Nelder-Mead simplex algorithm [12]. Finally, Fig. 2(c) also illustrates the other advantage of MM-game. Specifically, in contrast to MT-game, its trade-off frequency is continuous on the transmission power and so less sensitive to transmission power varying.

6 Conclusions

In this paper, we have suggested an anti-eavesdropping strategy maximizing minimal secrecy rate, or in other words, guaranteed (expected) secrecy rate strategy in the transmitter's communication with a group of nodes. The problem has been modeled and solved in a Bayesian game-theoretical framework between a transmitter and an adversary (eavesdropper) in the most dangerous scenario for the transmitter in which it does not know what type of antennas (directional or omni-directional) the adversary has. It has been proven that in contrast to maximizing total secrecy rate strategy, the suggested strategy supports uninterrupted communication with each node. Moreover, it leads to egalitarian communication. That is why, the suggested approach might give an extra advantage in scenarios in which mission success essentially depends on the reliable and secure coordination of each engaged node via communication.

Appendix

An Auxiliary Max-Min Problem: Let us find in Lemma 1 below a solution of the following associated auxiliary max-min problem:

$$\max\{\min_{i \in \mathcal{N}} a_i z_i : \mathbf{z} = (z_1, \ldots, z_n) \in \mathcal{P}\}, \tag{21}$$

where a_i, $i \in \mathcal{N}$ are non-negative constants.

Lemma 1. *(a) If there is an i such that $a_i = 0$ then $\min_{j \in \mathcal{N}} a_j z_j = 0$ for any $\mathbf{z} \in \mathcal{P}$, and so, any $\mathbf{z} \in \mathcal{P}$ is a solution of the max-min problem (21).*

(b) If $a_i > 0$ for all $i \in \mathcal{N}$ then the maxmin problem (21) has an unique solution and it is given as follows:

$$z_i = \frac{1/a_i}{\sum_{j \in \mathcal{N}}(1/a_j)} \text{ for } i \in \mathcal{N}. \tag{22}$$

Proof: (a) is obvious. Let $a_i > 0$ for all $i \in \mathcal{N}$. First let us prove that if $(z_1, \ldots, z_n)$ is a solution of the max-min problem (21), then

$$z_i a_i = z_j a_j \text{ for any } i \neq j. \tag{23}$$

Assume that (23) does not hold, and let $z_i a_i$ is the max-min value of (21). Since (23) does not hold, then there is a j such that

$$z_i a_i < z_j a_j. \tag{24}$$

Let $\mathcal{K} = \{k \in \mathcal{N} : a_k z_k = a_i z_i\}$ and K be the number of elements of this set. Then for enough small $\epsilon > 0$ we have that

$$\min_{k \in \mathcal{N}} a_k z_k = z_i a_i < \min_{k \in \mathcal{N}} a_k z_{k,\epsilon} \tag{25}$$

with

$$z_{k,\epsilon} = \begin{cases} z_k, & k \notin \mathcal{K} \cup \{j\}, \\ z_k + \epsilon/K, & k \in \mathcal{K}, \\ z_j - \epsilon, & k = j. \end{cases} \tag{26}$$

Noting that $(z_{1,\epsilon}, \ldots, z_{N,\epsilon}) \in \mathcal{P}$ for enough small positive ϵ, we obtain, by (25), that $(z_1, \ldots, z_n)$ cannot be a solution of the max-min problem (21). This contradiction proves that (23) holds. Then, there is a θ such that

$$a_k z_k = \theta, k \in \mathcal{N}. \tag{27}$$

So,

$$z_k = \theta/a_k, k \in \mathcal{N}. \tag{28}$$

Summing it over k implies that

$$1 = \sum_{k \in \mathcal{N}} z_k = \theta \sum_{k \in \mathcal{N}} (1/a_k). \tag{29}$$

Thus,

$$\theta = 1/\sum_{k \in \mathcal{N}} (1/a_k). \tag{30}$$

Substituting this θ into (28) implies (22). ∎

Proof of Theorem 1: By (8), the best response equation (11) is equivalent to max-min problem (21) with

$$a_i = T_i - \gamma E_i y_i - (1-\gamma) E_i^0. \tag{31}$$

By (4), (5) and (7), since $y_i \in [0,1]$ and $\gamma \in [0,1]$ we have that

$$T_i - \gamma E_i y_i - (1-\gamma) E_i^0 \geq \min\{T_i - E_i y_i, T_i - E_i^0\} > 0 \text{ for all } i \in \mathcal{N}. \tag{32}$$

Thus, $a_i > 0$ for any i. Then, by Lemma 1, the transmitter's best response $\boldsymbol{x} = (x_1, \ldots, x_n)$ is given as follows:

$$x_i = 1/(\theta(T_i - \gamma E_i y_i - (1-\gamma) E_i^0)) \tag{33}$$

with

$$\theta \triangleq \sum_{j \in \mathcal{N}} 1/(T_j - \gamma E_j y_j - (1-\gamma) E_j^0). \tag{34}$$

Since $y_i \in [0,1]$ and $\gamma \in [0,1]$, (33) and (34) with (4), (5) and (7) imply that $x_i > 0$ for $i \in \mathcal{N}$. Thus, without loss of generality we can assume that

$$x_i > 0 \text{ for any } i. \tag{35}$$

By (10), the best response equation (12) is equivalent to max-min problem (21) with

$$a_i = E_i x_i. \tag{36}$$

Moreover, by (35), $a_i > 0$ for all i. Then Lemma 1 implies that adversary's best response $\boldsymbol{y}$ to transmitter's strategy $\boldsymbol{x}$ is

$$y_i = 1/(E_i x_i \sum_{j \in \mathcal{N}} 1/(E_j x_j)) \text{ for } i \in \mathcal{N}, \tag{37}$$

which is equivalent to

$$y_i = 1/(E_i x_i \nu) \text{ for } i \in \mathcal{N} \text{ with } \nu = \sum_{i \in \mathcal{N}} 1/(E_i x_i). \tag{38}$$

Substituting (38) into (33) implies

$$x_i = 1/(\theta(T_i - \gamma/(x_i \nu) - (1-\gamma) E_i^0)). \tag{39}$$

Solving Eq.(39) on x_i implies

$$x_i = (1/\theta + \gamma/\nu)/(T_i - (1-\gamma) E_i^0). \tag{40}$$

Since $\boldsymbol{x} \in \mathcal{P}$ summing up (40) over $i \in \mathcal{N}$ implies

$$1 = (1/\theta + \gamma/\nu)\overline{X} \text{ with } \overline{X} \text{ given by (14)}. \tag{41}$$

Substituting (40) into (38) implies

$$y_i = (T_i - (1-\gamma)E_i^0)/(E_i(\nu/\theta + \gamma)) \text{ for } i \in \mathcal{N}. \qquad (42)$$

Summing up (42) over $i \in \mathcal{N}$, since $\boldsymbol{y} \in \mathcal{P}$, implies

$$1 = \overline{Y}/(\nu/\theta + \gamma) \text{ with } \overline{Y} \text{ given by (14)}. \qquad (43)$$

Multiplying (41) and (43) implies

$$1 = \overline{XY}/\nu. \qquad (44)$$

Thus,

$$\nu = \overline{XY}. \qquad (45)$$

Substituting this ν into (43) and solving on θ implies

$$\theta = \overline{XY}/(\overline{Y} - \gamma). \qquad (46)$$

Finally, substituting (33) into (8) implies that the transmitter's payoff for equilibrium strategies is $1/\theta$. Substituting (38) into (10) implies that the adversary's payoff for equilibrium strategies is $1/\nu$. These, (45) and (46) imply (15). ∎

References

1. Ahuja, B., Mishra, D., Bose, R.: Fair subcarrier allocation for securing OFDMA in IoT against full-duplex hybrid attacker. IEEE Trans. Inf. Forensics Secur. **16**, 2898–2911 (2021)
2. Altman, E., Avrachenkov, K., Garnaev, A.: Closed form solutions for water-filling problem in optimization and game frameworks. Telecommun. Syst. J. **47**, 153–164 (2011)
3. Boonyakarn, P., Komolkiti, P., Aswakul, C.: Game-based analysis of eavesdropping defense strategy in WMN with directional antenna. In: Proceedings of 8th International Joint Conference on Computer Science and Software Engineering (JCSSE), pp. 29–33 (2011)
4. Dai, H.N., Li, D., Wong, R.C.W.: Exploring security improvement of wireless networks with directional antennas. In: IEEE 36th Conference on Local Computer Networks, pp. 191–194 (2011)
5. Guo, D., Ding, H., Tang, L., Zhang, X., Yang, L., Liang, Y.C.: A proactive eavesdropping game in MIMO systems based on multiagent deep reinforcement learning. IEEE Trans. Wirel. Commun. **21**(11), 8889–8904 (2022)
6. Fudenberg, D., Tirole, J.: Game Theory. MIT Press, Boston, MA (1991)
7. Garnaev, A., Baykal-Gursoy, M., Poor, H.: A game theoretic analysis of secret and reliable communication with active and passive adversarial modes. IEEE Trans. Wirel. Commun. **15**, 2155–2163 (2016)
8. Garnaev, A., Trappe, W.: Bargaining over the fair trade-off between secrecy and throughput in OFDM communications. IEEE Trans. Inf. Forensics Secur. **12**, 242–251 (2017)

9. Garnaev, A., Trappe, W.: A sophisticated anti-eavesdropping strategy. IEEE Wirel. Commun. Lett. **11**, 1463–1467 (2022)
10. Garnaev, A., Trappe, W.: An eavesdropping and jamming dilemma with sophisticated players. ICT Express **9**(4), 691–696 (2023)
11. Han, Z., Niyato, D., Saad, W., Basar, T., Hjrungnes, A.: Game Theory in Wireless and Communication Networks: Theory, Models, and Applications. Cambridge University Press (2012)
12. Lagarias, J., Reeds, J., Wright, M., Wright, P.: Convergence properties of the Nelder–mead simplex method in low Dimensions. SIAM J. Optim. **9**(1), 112–147 (1998)
13. Li, X., Dai, H.N., Zhao, Q.: An analytical model on eavesdropping attacks in wireless networks. In: Proceedings of IEEE International Conference on Communication Systems, pp. 538–542 (2014)
14. Narbudowicz, A.: Multi-antenna system as an eavesdropper for directional beams. In: Proceedings of International Workshop on Antenna Technology (IWAT), pp. 1–4 (2020)
15. Yu, D.D., Cioffi, J.M.: Iterative water-filling for optimal resource allocation in OFDM multiple-access and broadcast channels. In: Proceedings of IEEE Globecom 2006, pp. 1–5 (2006)
16. Zhu, Q., Saad, W., Han, Z., Poor, H.V., Basar, T.: Eavesdropping and jamming in next-generation wireless networks: a game-theoretic approach. In: Proceedings of IEEE Military Communications Conference (MILCOM), pp. 119–124 (2011)

Economic Analysis of DMA-Constrained Data Sharing Strategies Among Platforms

Patrick Maillé[1](✉) and Bruno Tuffin[2]

[1] IMT Atlantique, IRISA, UMR CNRS 6074, Rennes, France
patrick.maille@imt.fr
[2] Inria, Univ. Rennes, CNRS, IRISA, Rennes, France

Abstract. To stimulate innovation and competition in the digital sector, the European Commission has adopted the Digital Markets Act (DMA), that aims at limiting the gatekeeping power of the biggest actors. For instance, such prevailing platforms may be forced to share with their competitors some, or all, of the data they collected on users that can help improve service quality.

This paper proposes a model to analyze the impacts of such a regulation, by comparing *(i)* a laisser-faire situation where prevailing platforms are free to sell data to newcomer service providers, *(ii)* the case when platforms can decide to share data but charging newcomers is prohibited, and *(iii)* a scenario where data sharing is imposed and enforced by the regulator. We investigate the impacts of those scenarios on key metrics like User Welfare, provider revenues, and user demand repartition.

1 Introduction

As a response to the digital sector being more and more controlled by a small number of (mostly US-based) giant companies, the European Commission newly adopted the Digital Markets Act (DMA), "the EU's law to make the markets in the digital sector fairer and more contestable" [3]. The law aims in particular at preventing actors with a powerful position, labeled *gatekeepers*, from stifling competition regarding services that they themselves propose. The gatekeeper position can indeed provide a significant advantage with respect to newcomer competitors, e.g., thanks to the access to user-specific information collected through the (often numerous) services those powerful actors offer.

The powerful companies targeted by the DMA are those satisfying three conditions: *(i)* a sufficiently significant size in terms of revenue, and an operation in at least three EU member states; *(ii)* a gateway position between businesses and consumers, concerning more than 45 million monthly active consumers and more than 10 thousand yearly active businesses established in the EU; and *(iii)* a durable such position, of at least three years.

For such actors, the DMA imposes some obligations, like data sharing with other businesses in order to prevent lock-in situations, and stimulate innovation and competition [2].

The objective of this paper is to provide a modeling methodology to investigate the influence of regulations based on those principles. More specifically, we consider a new service (say, app), for which two service providers compete to attract users. The first is a gatekeeper or incumbent platform, as those targeted by the DMA, that already has as customers (for its other services) a significant proportion of the target population of the new service; The other is a newcomer to the market, that therefore has no customer base but is highly innovative and could provide the new service with higher quality of experience.

Without any regulation, the incumbent already uses the user data collected to provide good service, hence attracting demand, and make revenue by introducing user-targeted ads within its service. By contrast, the newcomer would not benefit from that knowledge, hence a poorer service and/or less ad revenue.

To limit the advantage of the incumbent over potential competitors, the regulator can intervene with regard to that use of user personal data. For example, the use of user data collected through the incumbent platform can be regulated, i.e., limited to a certain range. The enforcement of GDPR (for General Data Protection Regulation) can be thought of this way (although its main goal is to let users control how their data is used) as it limits what platforms can do with user personal data.

Another possibility to level the playing field could be to impose the incumbent platform to share (some of) its user information, that the newcomer could use to improve its service (and get more demand) and/or make more revenue from better-targeted ads. This can be part of what the DMA intends to do.

Our goal is to investigate the data sharing decisions that are made, depending of who makes those decisions, and how they can be enforced. Then the consequences on demands, providers' revenues, and user welfare are considered. More specifically, we compare the following scenarios:

- The incumbent chooses the level of data sharing with the newcomer, but is not allowed to charge the newcomer for it;
- The incumbent is free to choose the data sharing level, and to charge the newcomer;
- The level of data sharing is imposed by the regulator, who then suffers some costs to enforce that policy.

Note that some contributions in the literature evoke the economic questions around the DMA and user data sharing [4,5], but to the best of our knowledge this is the first concrete mathematical modeling of the problem.

2 Modeling Data Sharing Among Providers

This section introduces the mathematical model we consider to represent how data sharing from an incumbent provider with a newcomer affects service qualities, and as a consequence user demands, provider revenues, and user welfare.

2.1 The Providers and Their Qualities

Throughout the whole paper, we use the index I (resp., N) to refer to the incumbent platform (resp., newcomer service provider). Those indices will also be used as labels to specify the choice made by users regarding the provider they choose for that service; finally we add the label "0" for users preferring not to use that service at all.

Let us denote by $\alpha \in [0,1]$ the proportion of user-related data controlled by the incumbent that the newcomer is able to access, and by $q_N(\alpha)$ the corresponding quality of the service that the newcomer can offer if accessing that amount of data, assumed non-decreasing. By contrast, the quality offered by the incumbent for the similar service is independent of α, and denoted by q_I. Finally, the quality experienced by users not considering any service is $q_0 = 0$.

We assume that the target population of the newcomer is included in the population already using some service(s) from the incumbent platform, so the incumbent already has the user-specific information, and the quality $q_N(\alpha)$ is considered the same for each potential user of the newcomer service. A larger α will make the newcomer more attractive, at the expense of the incumbent that will lose some users for that specific service (or get less traffic from users, who will partially turn to the newcomer).

We will often assume that any extra information can yield some quality improvement, but that there are diminishing returns, as formalized below.

Assumption A. *The quality function $q_N(\cdot)$ of the newcomer's service is differentiable, strictly increasing and concave over $[0,1]$.*

2.2 User Options and Choices

To represent the impact of quality on demands, we assume that each user i has a utility for each choice $C \in \{I, N, 0\}$ regarding the service (respectively, the incumbent, the newcomer, or no service), of the form

$$U_{i,C} = \ln(1 + q_C) + \kappa_{i,C}, \tag{1}$$

where q_C is the quality for the choice C, and $\kappa_{i,C}$ is a user-specific value. Each user then selects the highest-utility choice for them. Expression (1) can be justified as follows: the average perceived quality increasing logarithmically with quality is inspired from observations in psychophysics, regarding perception of a physical stimulus versus its magnitude [8]; and the user-specific terms account for all subjective aspects (limited rationality, reputation...) involved with evaluating an option.

Mathematically, when all $\kappa_{i,C}$ are independent random variables following a standard Gumbel distribution–an assumption often made in discrete-choice models [1]–then choices happen to spread proportionally to the exponentials of the non-random parts of the utilities (a result sometimes called the Gumbel-max trick). Making that assumption, the proportions of users choosing each option I, N, or 0 (for which we consider the quality to be zero) then respectively equal

$$d_I = \frac{1+q_I}{3+q_I+q_N(\alpha)}, \quad d_N = \frac{1+q_N(\alpha)}{3+q_I+q_N(\alpha)}, \quad d_0 = \frac{1}{3+q_I+q_N(\alpha)}.$$

2.3 Revenues for Providers

We assume both service providers make money through ads, extracting some revenue that is proportional to their market share for the service.

Let us denote by β_I (resp., β_N) the revenue per unit of market share that the incumbent platform (resp., the newcomer) can generate through the service. Their total revenue R_I (resp., R_N) can then be expressed as a function of α:

$$R_I = \beta_I \frac{1+q_I}{3+q_I+q_N(\alpha)} \quad \text{and} \quad R_N = \beta_N \frac{1+q_N(\alpha)}{3+q_I+q_N(\alpha)}. \tag{2}$$

Note that we can also consider β_N to be a non-decreasing function of α, as more targeted ads, which we assumed are positively perceived by users, are also more likely to lead to sales and generate revenue.

2.4 User Welfare

Intuitively, User Welfare (UW) should be maximized when α is maximal, since α just increases the quality of Choice N without affecting the others. To quantify UW, we use as a reference the situation where no service is provided (and each user i has a utility $U_{i,0}$), and define UW as the average gain in utility with respect to that reference. As each user i selects the highest-utility choice, their gain G_i from the existence of the service is

$$G_i = \max(U_{i,I}, U_{i,N}, U_{i,0}) - U_{i,0} = \max(U_{i,I} - U_{i,0}, U_{i,N} - U_{i,0}, 0).$$

Taking the average over all users (hence removing the user index), we get

$$\text{UW} = \mathbb{E}[G] = \int_{x=0}^{\infty} \mathbb{P}(G > x) \mathrm{d}x = \int_{x=0}^{\infty} (1 - \mathbb{P}(G \leq x)) \, \mathrm{d}x.$$

Now, remark that $G = \max(\ln(1+q_I) + \kappa_I - \kappa_0, \ln(1+q_N(\alpha)) + \kappa_N - \kappa_0, 0)$, so that for any $x \geq 0$ we have

$$\{G \leq x\} = \{\kappa_0 + x \geq \ln(1+q_I) + \kappa_I\} \cap \{\kappa_0 + x \geq \ln(1+q_N(\alpha)) + \kappa_N\}.$$

Hence $\mathbb{P}(G \leq x)$ is the probability that a user would prefer the "no service" option if that option had a non-random part of x instead of 0. Applying again results from discrete-choice theory [1] gives $\mathbb{P}(G \leq x) = \frac{e^x}{e^x + 2 + q_I + q_N(\alpha)}$, thus

$$\text{UW} = \int_{x=0}^{\infty} \frac{2+q_I+q_N(\alpha)}{e^x + 2 + q_I + q_N(\alpha)} \mathrm{d}x = \ln(3+q_I+q_N(\alpha)). \tag{3}$$

As expected, UW increases with α, hence a regulator focusing only on that metric would want to impose a maximum level of data sharing among service providers, as the DMA tries to enforce.

3 When Information Sharing Is Decided by the Incumbent Platform

In this section, we investigate the level of information sharing α that would be chosen, if the decision was taken by the incumbent platform trying to maximize its revenue.

3.1 The Incumbent Will Not Share Data Without Compensation or Regulation

We first assume that the incumbent platform can choose the level of data sharing with the newcomer, but that it is not allowed to ask for any (financial) compensation. In that simple case, the incumbent revenue R_I in (2) being decreasing in α, the optimal sharing level is just $\alpha = 0$. Note that this situation is the worst from the point of view of User Welfare, as pointed out before.

3.2 An All-or-Nothing Outcome if the Incumbent Can Charge the Newcomer

Now assume that no regulation at all applies, so that the incumbent is free to decide the data sharing level α but also to charge the newcomer provider any price it decides to.

In such a situation, the incumbent, having the decision power over α, can extract all the potential benefits in ad revenues stemming from data sharing. To do so, it would charge the newcomer just so that the newcomer's net revenue is equal (or very slightly above, to introduce some minor incentive) what it would get with no user data at all.

The objective for the incumbent would then be to maximize the *sum* of both providers' revenues, namely

$$R_I + R_N = \beta_I \frac{1 + q_I}{3 + q_I + q_N(\alpha)} + \beta_N \frac{1 + q_N(\alpha)}{3 + q_I + q_N(\alpha)}$$
$$= \beta_N \frac{q_N(\alpha) + 1 + \frac{\beta_I}{\beta_N}(1 + q_I)}{q_N(\alpha) + 3 + q_I}.$$

That expression is strictly increasing in $q_N(\alpha)$, if and only if $3 + q_I > 1 + \frac{\beta_I}{\beta_N}(1 + q_I)$, i.e., if $\frac{\beta_N}{\beta_I} > \frac{1+q_I}{2+q_I}$, that is, if the newcomer ad profitability relative to the incumbent's is high enough with respect to a function of the incumbent quality.

When that inequality holds, the pair *incumbent+newcomer* generates the most revenue when $\alpha = 1$ (which is the only optimal solution when $q_N(\cdot)$ is strictly increasing). The extra revenue R_{extra} with respect to the situation where $\alpha = 0$ is

$$R_{\text{extra}} = \frac{\beta_I(1 + q_I) + \beta_N(1 + q_N(1))}{3 + q_I + q_N(1)} - \frac{\beta_I(1 + q_I) + \beta_N(1 + q_N(0))}{3 + q_I + q_N(0)}.$$

To keep all this benefit to itself, the incumbent could present the newcomer with a take-it-or-leave-it offer, adjusting the price p it charges the newcomer to have it exactly indifferent between accepting and refusing, i.e.,

$$\beta_N \frac{1+q_N(1)}{3+q_I+q_N(1)} - p = \beta_N \frac{1+q_N(0)}{3+q_I+q_N(0)}.$$

Alternatively, the newcomer could be aware that it has some bargaining power, as the benefit only materializes when the newcomer accepts the offer, and could choose to refuse such an offer. The resulting outcome, seen either as the Nash Bargaining Solution [7] with equal bargaining power, or a Shapley value [6], would then consist in both providers sharing that benefit evenly. In that case the price $\bar{p}$ paid by the newcomer to the incumbent would be such that

$$\beta_N \frac{1+q_N(1)}{3+q_I+q_N(1)} - \bar{p} = \beta_N \frac{1+q_N(0)}{3+q_I+q_N(0)} + \frac{1}{2} R_{\text{extra}}.$$

Summarizing, the total utility (revenues from ads plus monetary exchanges between providers) of the incumbent and newcomer providers are respectively:

With no bargaining: $\begin{cases} U_I = \beta_I d_I|_{\alpha=0} + R_{\text{extra}} \\ U_N = \beta_N d_N|_{\alpha=0} \end{cases}$

At Nash Bargaining Solution (NBS): $\begin{cases} U_I = \beta_I d_I|_{\alpha=0} + R_{\text{extra}}/2 \\ U_N = \beta_N d_N|_{\alpha=0} + R_{\text{extra}}/2. \end{cases}$

Whether the incumbent has all the negotiating power or that power is shared so that the gain R_{extra} is equally split over providers, the data sharing level α_I^* that should be chosen by the incumbent (hence the index I) is one maximizing R_{extra}, or equivalently $R_I + R_N$. Hence the following result.

Proposition 1. *Under Assumption A, the data sharing level α_I^* that the incumbent would choose if allowed to charge the newcomer, is*

$$\alpha_I^* = \begin{cases} 1 & \text{if } \frac{\beta_N}{\beta_I} > \frac{1+q_I}{2+q_I} \\ 0 & \text{if } \frac{\beta_N}{\beta_I} < \frac{1+q_I}{2+q_I} \\ \text{any } \alpha \in [0,1] & \text{if } \frac{\beta_N}{\beta_I} = \frac{1+q_I}{2+q_I}. \end{cases} \quad (4)$$

Equation (4) states that if the newcomer's ads are sufficiently efficient (in terms of revenue β_N per market share unit) then it is worth maximizing the newcomer's quality to maximize total revenue. While this could be expected, the condition also specifies the threshold for data sharing to be beneficial, and surprisingly it is not even necessary to have $\beta_N > \beta_I$, although that would be sufficient: Even with equally effective advertising, i.e., $\beta_N = \beta_I$, it is optimal to select $\alpha = 1$.

Conversely, if $\frac{\beta_N}{\beta_I} < \frac{1+q_I}{2+q_I}$ then the incumbent's best strategy is to set $\alpha = 0$, i.e., share no user data with the newcomer. Indeed, for any positive α there would be no price the incumbent could charge the newcomer that would at the same time be accepted by the newcomer and lead to some gain (between ad revenue and selling data) to the incumbent.

4 When Information Sharing is Enforced by the Regulator

We already saw from (3) that User Welfare strictly increases with $q_N(\alpha)$, so a regulator focusing on UW only would want to impose the incumbent a maximum level of data sharing, without any payment from the newcomer to the incumbent, in order to stimulate innovation and competition. But that is assuming that enforcing such a rule is costless, while in practice it might be complex. In this section, we add a regulation cost component to the objective function of the regulator, to highlight the possible loss of welfare such a cost incurs.

4.1 Modeling the Cost of Enforcing Data Sharing

First note that for some basic pieces of data (e.g., user name or address) it may be relatively easy to determine whether they have been shared truthfully, for example the recipient (the newcomer provider) can check that information. However, more detailed information, like what the incumbent exactly learned about users from their online behavior, can be altered by the incumbent to prevent its competitor from maximizing service quality, or can simply be hidden by the incumbent, the newcomer not even knowing that such information exists. It can then be extremely difficult to identify missing or deformed data.

Therefore, ordering data by increasing checking costs, we end up with a cost function $c(\alpha)$ to enforce a sharing level α satisfying the assumption below (where we also add differentiability).

Assumption B. *The cost function $c(\cdot)$ of enforcing some level of data sharing is differentiable, strictly increasing and convex.*

4.2 Optimizing the Regulated Data Sharing Level

The regulator is now faced with a trade-off regarding the data sharing level α to set, a higher α meaning larger User Welfare from (3) but also higher enforcing costs. We formulate its objective function, or utility U_R, as

$$U_R(\alpha) = \mathrm{UW}(\alpha) - c(\alpha) = \ln(3 + q_I + q_N(\alpha)) - c(\alpha).$$

Maximizing $U_R(\alpha)$ is a convex optimization problem under Assumptions A and B, in which case the first-order condition

$$\frac{q'_N(\alpha)}{3 + q_I + q_N(\alpha)} = c'(\alpha), \qquad (5)$$

has at most one solution. Therefore we have the following result.

Proposition 2. *The maximization problem for the regulator $\max_{\alpha \in [0,1]} U_R(\alpha)$ has a unique solution α_R^*, that is*

$$\alpha_R^* = \begin{cases} 0 \text{ if } c'(0) > \frac{q'_N(0)}{3+q_I+q_N(0)}, \\ 1 \text{ if } c'(1) < \frac{q'_N(1)}{3+q_I+q_N(1)}, \\ \text{the only value } x \in [0,1] \text{ such that } \frac{q'_N(x)}{3+q_I+q_N(x)} = c'(x) \text{ otherwise.} \end{cases}$$

The first and second cases in the proposition above respectively correspond to situations where controls are just too expensive to impose any level of data sharing, and where on the opposite those costs are negligible with respect to the gains in User Welfare.

5 A Numerical Analysis

In this section, we provide a quantitative analysis for some specific functions $q_N(\cdot)$ and $c(\cdot)$. While any functions satisfying Assumptions A and B are easy to deal with numerically (as (5) can be solved by dichotomy), we take here some functions allowing to solve (5) analytically, namely we consider quadratic checking costs for the regulator, and a service quality for the newcomer that increases linearly with the level of data sharing:

$$q_N(\alpha) = \gamma + M\alpha \tag{6}$$
$$c(\alpha) = K\alpha^2 \tag{7}$$

for some constants $\gamma \geq 0$, $M > 0$, and $K > 0$.

With those functions, the first-order condition (5) of the regulator's problem leads to the second-degree equation $\alpha^2 + \frac{3+q_I+\gamma}{M}\alpha - \frac{1}{2K} = 0$, which has a unique non-negative solution. As per Proposition 2, the sharing level imposed by the regulator would then be $\alpha_R^* = \min\left(1, \sqrt{\left(\frac{3+q_I+\gamma}{2M}\right)^2 + \frac{1}{2K}} - \frac{3+q_I+\gamma}{2M}\right)$.

On the other hand, note that Proposition 1 implies that the data sharing level, if chosen by the incumbent, is independent of the functions we consider, as it only depends on ad efficiencies β_I, β_N and incumbent service quality q_I.

In Fig. 1, we vary individual parameters of the model, keeping the other parameters to the default values below:

β_I	β_N	q_I	γ	M	K
1	1	1	0.5	1	0.5

The results are displayed in Fig. 1. The plots illustrate the threshold effect for α_I^* regarding newcomer ad profitability β_N: below a threshold (here 2/3), an incumbent allowed to charge the newcomer prefers not to share anything, while it would share all of its data above that threshold, even if some bargaining takes place between both providers. Without bargaining, the incumbent takes all the benefits of the sharing, while this benefit is shared equally with the provider in the Nash Bargaining Solution outcome.

As could be expected, from the incumbent's point of view, the preferred situation is the "no bargaining" outcome, then the bargaining one, then the situation with no sharing ($\alpha = 0$), its least-preferred being the regulated scenario, where it is forced to share some amount α_R^* of data without any compensation. But for the newcomer, while the "no bargaining" solution yields the lowest utility, exactly as the "$\alpha = 0$" situation, the best scenario can either be the Nash Bargaining Solution or the regulated one, the latter being preferred for high

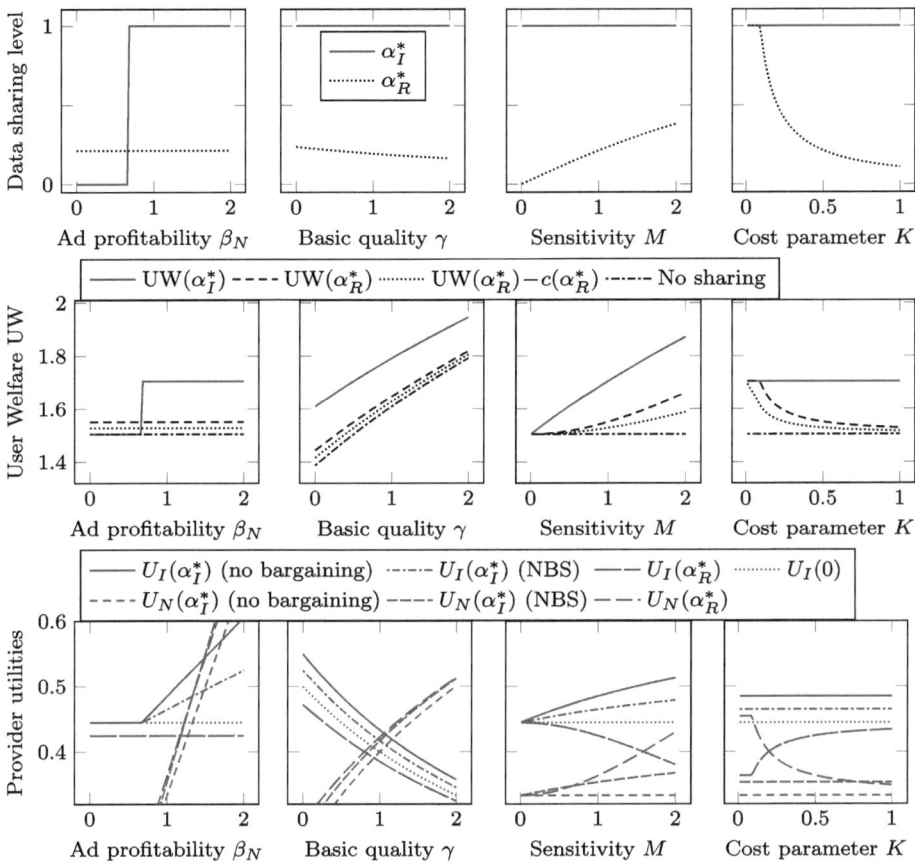

Fig. 1. Impact of model parameters on the selected data sharing level α *(top)*, User Welfare *(center)*, and providers' utilities *(bottom)*.

values of K (regulation is so costly that the imposed sharing level is insufficient), or low values of M where the imposed sharing level is also small, as the curves of α_R^* show.

In terms of User Welfare (or UW minus the regulation cost), the best situation can either be the laisser-faire (i.e., letting the incumbent decide the sharing level, with or without bargaining) when $\alpha_I^* = 1$, which can be at the expense of the newcomer, or some regulatory intervention even if imposing α_R^* is costly.

6 Conclusions

This paper proposes a simple mathematical model to analyze the economic implications of regulating the sharing of user data that powerful platforms control, one key component of the DMA. The model encompasses competition between incumbent and newcomer actors, and the feasibility of regulation. Our results suggest that while preventing payments from newcomers to incumbents is simple to implement, it would not lead to data sharing or stimulate innovation by helping newcomers if data sharing is not imposed. Then, facing the reality of enforcing such regulation leads to tradeoffs about what level of sharing can be imposed: in the end, from a user welfare point of view, or even for a newcomer with bargaining power, a regulatory intervention is not always beneficial.

The model presented in this paper can be enriched in several dimensions, for example by considering more complex relations between data sharing and quality, with data sharing not being limited to one number, or by introducing different demand models, possibly with several newcomers.

References

1. Ben-Akiva, M., Lerman, S.: Discrete Choice Analysis. MIT Press, Cambridge (1985)
2. Centre for Information Policy Leadership. Data Sharing Obligations Under the DMA: Challenges and Opportunities. CIPL Discussion Paper (2024)
3. European Commission. Regulation (EU) 2022/1925 of the European Parliament and of the Council of 14 September 2022 on contestable and fair markets in the digital sector and amending Directives (EU) 2019/1937 and (EU) 2020/1828 (Digital Markets Act). Document 32022R1925 (2022)
4. Graef, I., Prüfer, J.: Governance of data sharing: a law & economics proposal. Res. Pol. **50**(9) (2021)
5. Martens, B., De Streel, A., Graef, I., Tombal, T., Duch Brown, N.: Business-to-business data sharing: An economic and legal analysis. Technical report, JRC Digital Economy Working Paper (2020)
6. Moulin, H.: Fair Division and Collective Welfare. MIT Press, Cambridge (2004)
7. Osborne, M., Rubinstein, A.: A Course in Game Theory. MIT Press, Cambridge (1994)
8. Reichl, P., Schatz, R., Tuffin, B.: Logarithmic laws in service quality perception: where microeconomics meets psychophysics and quality of experience. Telecommun. Syst. **52**(2), 587–600 (2013)

Achieving a Collective Target Through Incentives

K. S. Ashok Krishnan[1,2](✉), Hélène Le Cadre[3], and Ana Bušić[1,2]

[1] Inria, Paris, France
ana.busic@inria.fr
[2] DI ENS, École Normale Supérieure, PSL Research University, Paris, France
ashok-krishnan.komalan-sindhu@inria.fr
[3] Inria, University of Lille, CNRS, Centrale Lille, UMR 9189 CRIStAL, 59000 Lille, France
helene.le-cadre@inria.fr

Abstract. In many games, the payoff of individual users is a function of a collective outcome that is common to all agents. Often, the users may be interested in jointly steering this outcome to a desired value. This work presents such a scenario, where the collective outcome is strictly monotone in the joint strategy of the agents. Further, the agents are irrational in their perception of a random cost component in their payoff. This irrationality is modelled using prospect theory. A coordinator steers the game to a desired collective outcome, by designing incentives. These incentives modify the responses of the users. Owing to the potential structure of the game, the system converges to a Nash equilibrium at which the desired collective outcome is obtained.

Keywords: Game Theory · Equilibrium · Incentives · Prospect Theory

1 Introduction

In many game scenarios, the payoff of individual agents depends on some collective outcomes that are functions on the joint strategy of all agents [9]. In such games, the players themselves, or a game coordinator, may be interested in steering the game to reach a certain solution [6]. This may be needed to achieve some performance metric, such as economic efficiency. This steering is even more important when agents no longer behave rationally. Game theory is built on the assumption that agents are rational, performing the expected utility maximization [20]. However, humans are known to be non rational in their decision making, especially under uncertainty [8]. There are a number of models that incorporate different aspects of bounded rationality of humans playing a game [1]. Prospect theory (PT) [10] is one such model, which incorporates elements of behavioural economics, psychology and decision theory. Prospect theory can be used to capture the variations of human decision making under risky prospects.

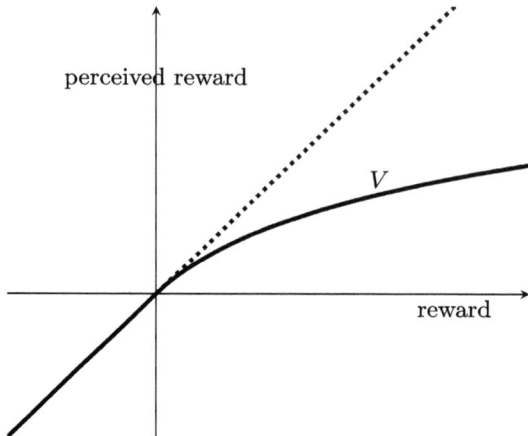

Fig. 1. Example of a prospect theoretic value function V that maps rewards to perceptions. The dotted line is the unit slope line for reference.

In this paper, prospect theoretic models are applied to a noncooperative game, where the payoff of the agents depends on a collective usage benefit. In [12], we proved that the presence of bounded rational agents changes the shape of the set of equilibria solutions of the expected utility maximization game. In the current paper, a coordinator steers the evolution of the game by designing incentives. The aim of this control is to steer the equilibrium to a point, at which the collective usage benefit reaches a desired value. Thus, the incentive compensates for the deviation in equilibrium caused by irrationality, while guaranteeing the reaching of a collective target.

1.1 Overview of Prospect Theory

In this section, we explain the PT framework, using ideas from works such as [10, 18]. An agent obtains random rewards $(R_1, ..., R_M)$ with probabilities $(q_1, ..., q_M)$ where $q_1 + ... + q_M = 1$. We call $(R_1, q_1, ..., R_M, q_M)$ a *prospect*. Assume that the rewards are ordered in increasing order of attractiveness. Now consider an agent having to choose between two prospects, $P := (R_1, q_1, ..., R_M, q_M)$ and $\hat{P} := (\hat{R}_1, \hat{q}_1, ..., \hat{R}_M, \hat{q}_M)$. This choice is modelled by prospect theory as follows. Each reward is viewed with respect to a (psychological) reference value. A reward which is greater than this reference is perceived as a gain, and a reward below this reference is understood as a loss. When faced with gains, agents will behave in a *risk averse* manner, and when faced with losses, agents tend to be *risk seeking* or *risk neutral* [10]. This perception is captured by a value function $V : \mathbb{R} \to \mathbb{R}$, as in Fig. 1. Here the agents are risk averse for gains and risk neutral for losses, with the reward zero representing the reference value that demarcates gains from losses. In addition to V, the probabilistic weight accorded to rewards is modified using a function π. The value of the prospect under PT is,

$$V = \sum_{j=1}^{M} \tilde{q}_j V(R_j), \qquad (1)$$

where $\tilde{q}_i$ is given by [18],

$$\tilde{q}_1 = \pi(q_1), \ \tilde{q}_j = \pi(\sum_{m=1}^{j} q_m) - \pi(\sum_{m=1}^{j-1} q_m), \ j = 2, ..., M, \qquad (2)$$

for some monotone increasing $\pi : [0,1] \to [0,1]$. The map π achieves the overweighting of small probabilities, and the underweighting of large probabilities, for large outcomes. Faced with a choice between two prospects, the agent compares their PT values (1), and chooses the better. Note that with $V(x) = x$ and $\pi(x) = x$, we retrieve the standard formulation of choice under the maximization of expected utilities.

1.2 Literature Review

Prospect theory has been used in economics for many decades, in order to model decision making under uncertainty [7]. Some works that use PT models in finite games include [11,13,14,19]. These works provide results for existence of Nash equilibria. In [17], a piecewise linear PT transformation is considered, and results are obtained for existence of equilibria. A broad survey of different papers that have used prospect theory in modelling economics of power systems is presented in [5]. In [4] existence results for local Nash equilibria are obtained under non smooth PT transformations. In [12], the effect of PT transforms on the set of equilibria is characterized analytically, and the results applied to an electricity market.

1.3 Contributions

In this paper, we show how incentives can be used to steer a game to a desired outcome, even in the presence of irrational agents. We present conditions under which such a steering algorithm converges to the target collective usage benefit. For specific forms of the utility function, we show how the Nash equilibria can be computed numerically using the fixed point of a function and also demonstrate the monotonicity of the Nash equilibrium strategy with the incentive parameter.

2 The Game Model

We consider a game with $N+1$ players. Let $\mathcal{N} := \{1, ..., N+1\}$ denote the set of players. Of these, player $N+1$ acts as a coordinator between the other N players. The coordinator chooses its strategy vector $\mathbf{p} = (p_1, ..., p_N) \in \mathcal{P} \subseteq \mathbb{R}^N$. For $i = 1, .., N$, player i chooses its strategy x_i from a compact set $\mathcal{X}_i \subset$

R. Let $\mathcal{X} := \prod_{j=1}^{N} \mathcal{X}_j$ denote the set of joint strategies. The randomness in outcomes is modelled by a real valued random variable ξ taking values over the set $\Xi := \{\xi_1, ..., \xi_M\}^1$ where $0 < \xi_1 < \cdots < \xi_M$, with probability distribution $\mathbf{q} = (q_1, ..., q_M)$. Let $V_i : \mathbb{R} \to \mathbb{R}, i = 1, ..., N$ and $\pi : [0,1] \to [0,1]$ be monotone increasing. The utility for user i playing x_i, when the other players play the joint strategy $\mathbf{x}_{-i} = (x_1, .., x_{i-1}, x_{i+1}, ..., x_N)$, is given by

$$J_i^{V_i, \pi}(x_i, \mathbf{x}_{-i}, \mathbf{p}) = a_i \mathcal{J}(\mathbf{x}) + \mathbb{E}_{\tilde{\mathbf{q}}}[V_i \circ \mathcal{R}_i(x_i, \xi)] - p_i x_i, \qquad (3)$$

where $a_i > 0$ for all i, and $\tilde{\mathbf{q}}$ is the distribution generated from $\mathbf{q}$ by (2) using π. This form of the utility function includes the following terms:

1. A collective usage benefit $\mathcal{J}(\mathbf{x}) : \mathcal{X} \to \mathbb{R}$, common to all agents.
2. An individual random reward $\mathcal{R}_i(x_i, \xi) : \mathcal{X}_i \times \Xi \to \mathbb{R}$. This reward is random due to its dependence on ξ, and is viewed subjectively through the PT value function V_i.
3. A fixed charge $p_i x_i$. Agent i is charged at the fixed rate p_i, for unit x_i. This term can be interpreted as an incentive for the agent to use a particular strategy.

This models a coordination game with agent specific costs and charges.

The PT game $\mathcal{G}_{PT}$ is the tuple $(\mathcal{N}, \mathcal{P}, \{\mathcal{X}_j\}_{j=1}^N, \{J_j^{V_j, \pi}\}_{j=1}^N, \{V_j\}_{j=1}^N, \pi)$. We also define the EUT game $\mathcal{G}_{EUT}$ as the instance of the PT game $\mathcal{G}_{PT}$ with $V_i(x) = x$ for $i = 1, ..., N$ and $\pi(x) = x$. $\mathcal{G}_{EUT}$ corresponds to the case of rational players working with expected utility maximization.

A Nash equilibrium for such a game, given $\mathbf{p} \in \mathcal{P}$, is any $\mathbf{x}^* = (x_i^*, \mathbf{x}_{-i}^*) = (x_1^*, ..., x_N^*) \in \mathcal{X}$ such that for all $i = 1, ..., N$,

$$J_i^{V_i, \pi}(x_i^*, \mathbf{x}_{-i}^*, \mathbf{p}) \geq J_i^{V_i, \pi}(x, \mathbf{x}_{-i}^* \mathbf{p}), \; \forall x \in \mathcal{X}_i. \qquad (4)$$

2.1 Existence, Uniqueness of Nash Equilibrium, and Potential Structure of the Game

We make the following assumptions on the functions constituting the utility function (3). These are sufficient conditions to ensure existence and uniqueness of Nash equilibria, and are commonly used in the literature.

Assumption 1. *$\mathcal{J}(\boldsymbol{x})$ is concave and differentiable with gradient $\nabla \mathcal{J}(\boldsymbol{x})$, which satisfies the strict monotone property,*

$$\langle \boldsymbol{x} - \boldsymbol{y}, \nabla \mathcal{J}(\boldsymbol{x}) - \nabla \mathcal{J}(\boldsymbol{y}) \rangle < 0 \text{ for all distinct } \boldsymbol{x}, \boldsymbol{y} \in \mathcal{X}.$$

Assumption 2. *$\mathcal{R}_i(x_i, \xi)$ is linear and increasing in x_i, for each i.*

---

[1] Most of the results in this work can be extended to the case where ξ has continuous distribution, albeit with more conditions on other variables to ensure integrability. We restrict ourselves to a finite support for simplicity.

Assumption 3. V_i is a concave, non decreasing and differentiable function for each i.

Theorem 1. *For any $\boldsymbol{p} \in \mathcal{P}$, there exists a unique Nash equilibrium for $\mathcal{G}_{PT}$.*

Proof. From Assumptions 1-3, it follows that for each i, $J_i^{V_i,\pi}(x_i,\cdot,\cdot)$ is concave in x_i. Thus $\mathcal{G}_{PT}$ is a concave game. Hence the game has at least one Nash equilibrium [16, Theorem 1]. Now we show that the game satisfies Rosen's diagonal strict concavity condition. Let $\mathbf{s} = (\frac{1}{a_1}, \ldots, \frac{1}{a_N})$, and let $\nabla_i J_i := \frac{\partial J_i^{V_i,\pi}(x_i,\mathbf{x}_{-i},\mathbf{p})}{\partial x_i}$. Define the pseudogradient,

$$v_\mathbf{s}(\mathbf{x}) = [s_1 \nabla_1 J_1, \cdots, s_N \nabla_N J_N]. \quad (5)$$

Then we have,

$$\langle \mathbf{x} - \hat{\mathbf{x}}, v_\mathbf{s}(\hat{\mathbf{x}}) - v_\mathbf{s}(\mathbf{x}) \rangle = \langle \mathbf{x} - \hat{\mathbf{x}}, \nabla \mathcal{J}(\hat{\mathbf{x}}) - \nabla \mathcal{J}(\mathbf{x}) \rangle$$
$$+ \sum_{i=1}^N \left(\frac{x_i - \hat{x}_i}{a_i}\right) \mathbb{E}_{\tilde{\mathbf{q}}}[r_i(\xi)(V_i' \circ \mathcal{R}_i(\hat{x}_i, \xi) - V_i' \circ \mathcal{R}_i(x_i, \xi))],$$

where V_i' indicates the derivative of V_i and $r_i(\xi) = \frac{\partial \mathcal{R}_i(x_i,\xi)}{\partial x_i}$. The first term in the above sum is strictly positive from Assumption 1. Since V_i is concave (Assumption 3), $V_i'(x)$ is decreasing in x. Using Assumption 2 with this, it follows that the second term in the above sum is non negative. Hence $\langle \mathbf{x} - \hat{\mathbf{x}}, v_\mathbf{s}(\hat{\mathbf{x}}) - v_\mathbf{s}(\mathbf{x}) \rangle > 0$ for all distinct $\mathbf{x}, \hat{\mathbf{x}}$. Hence, $\mathcal{G}_{PT}$ satisfies the diagonal strict concavity condition, from [16, Theorem 2] the Nash equilibrium given $\mathbf{p}$ is unique.

Corollary 2. *The game $\mathcal{G}_{EUT}$ has a unique Nash equilibrium, at each $\boldsymbol{p}$.*

It is also easy to see that $\mathcal{G}_{PT}$ is a weighted potential game [15]; we will rely on this property to design the steering algorithm in Sect. 3.

Lemma 3. *The game $\mathcal{G}_{PT}$ is a weighted potential game, with potential function*

$$\Phi(\boldsymbol{x}) := \mathcal{J}(\boldsymbol{x}) + \sum_{j=1}^N \frac{1}{a_j} \mathbb{E}_{\tilde{\mathbf{q}}}[V_j \circ \mathcal{R}_j(x_j, \xi) - p_j x_j]. \quad (6)$$

Proof. For all i, observe that for $\mathbf{x} = (x_1, \ldots, x_N) = (x_i, \mathbf{x}_{-i})$ and $\tilde{\mathbf{x}} = (\tilde{x}_i, \mathbf{x}_{-i})$, we have

$$J_i^{V_i,\pi}(x_i, \mathbf{x}_{-i}, \mathbf{p}) - J_i^{V_i,\pi}(\tilde{x}_i, \mathbf{x}_{-i}, \mathbf{p}) = a_i(\Phi(\mathbf{x}) - \Phi(\tilde{\mathbf{x}})). \quad (7)$$

3 Using Incentives to Steer the Collective Usage Benefit to a Desired Value

From the structure of the game and the equilibrium solution in Sect. 2, we observe that the equilibria are parametrized by the incentive vector $\mathbf{p}$. The game coordinator can steer the value of the collective benefit $\mathcal{J}(\mathbf{x})$ in a desired direction by choosing the incentives appropriately. Let α be a target value for $\mathcal{J}(\mathbf{x})$, and let $\hat{\mathbf{x}}$ be such that $\mathcal{J}(\hat{\mathbf{x}}) = \alpha$. The *inverse control problem* for the coordinator, when it has full information about the agents and their utilities, is to drive the system to $\hat{\mathbf{x}}$, by computing the parameter $\hat{\mathbf{p}}$ which satisfies

$$\frac{\partial J_i^{V_i,\pi}(p_i, x_i, \mathbf{x}_{-i})}{\partial x_i}\bigg|_{\mathbf{x}=\hat{\mathbf{x}}, \mathbf{p}=\hat{\mathbf{p}}} = 0, \qquad (8)$$

and then generate a sequence $\mathbf{p}(n)$ that converges to $\hat{\mathbf{p}}$. An $\hat{\mathbf{x}} \in \mathcal{X}$ is called reachable, if there exists $\hat{\mathbf{p}} \in \mathcal{P}$ such that

$$\hat{p}_i = a_i \frac{\partial \mathcal{J}(\mathbf{x})}{\partial x_i}\bigg|_{\mathbf{x}=\hat{\mathbf{x}}} + \mathbb{E}_{\tilde{q}}[r_i(\xi) V_i'(\mathcal{R}_i(\hat{x}_i, \xi))].$$

If $\hat{\mathbf{x}}$ is reachable by an incentive $\hat{\mathbf{p}}$, then $\hat{\mathbf{x}}$ is the Nash equilibrium of $\mathcal{G}_{PT}$ at $\hat{\mathbf{p}}$.

Lemma 4. *Let α be such that $\mathcal{J}(\hat{\mathbf{x}}) = \alpha$ at $\hat{\mathbf{x}}$ which is achievable at some $\hat{\mathbf{p}}$. Let the agents update their strategies by gradient descent,*

$$x_i(n+1) = x_i(n) + \Delta(n) \frac{\partial J_i^{V_i,\pi}(x_i, \mathbf{x}_{-i}, \mathbf{p}(n))}{\partial x_i}, \qquad (9)$$

where $\mathbf{p}(n) \to \hat{\mathbf{p}}$ such that $\|\mathbf{p}(n) - \hat{\mathbf{p}}\| \leq \Delta(n) C$ with $C \in \mathbb{R}_+^\star$, $\Delta(n)$ satisfies $\sum_{n=1}^{\infty} \Delta(n) = \infty$ and $\sum_{n=1}^{\infty} \Delta^2(n) < \infty$. Then $\mathbf{x}(n) \to \hat{\mathbf{x}}$ as $n \to \infty$.

Proof. Observe that (9) is equivalent to

$$x_i(n+1) = x_i(n) + \Delta(n) \left(\frac{\partial J_i^{V_i,\pi}(x_i, \mathbf{x}_{-i}, \hat{\mathbf{p}})}{\partial x_i} + \hat{p}_i - p_i(n) \right),$$

which is a gradient descent with decreasing error. Under our assumptions, it follows [2] that this gradient descent will lead to (agent wise) gradient descent on the functions $J_i^{V_i,\pi}(x_i, \mathbf{x}_{-i}, \hat{\mathbf{p}})$. Since the game $\mathcal{G}_{PT}$ is a weighted potential game, it follows that this gradient descent will converge to the Nash equilibrium of the game $\mathcal{G}_{PT}$ at parameter $\hat{\mathbf{p}}$ [3], which is the point $\hat{\mathbf{x}}$.

4 Balancing Collective and Individual Targets

To illustrate the PT game described in the preceding sections, consider an example of a collective of users buying electricity from an electricity producer. Each user buys x_i units from the producer at a price p_i per unit. Each user has a consumption threshold d_i. This could, for example, represent a minimum consumption level to meet their requirements. If the purchased quantity is above this threshold, then the user benefits from the purchase; otherwise it is a loss. The individual random reward is

$$\mathcal{R}_i(x_i, \xi) = (x_i - d_i)\xi, \tag{10}$$

where ξ models randomness of outcome. The collective usage benefit is

$$\mathcal{J}(\mathbf{x}) = -\sum_{i=1}^{N}(x_i - d)^2, \quad \mathcal{R}_i(x_i, \xi) = (x_i - d_i)\xi. \tag{11}$$

This represents an attempt by the collective of the users to keep the purchase vector $\mathbf{x}$ close to some desired vector $(d, ..., d)$. Here d represents a target energy consumption, for example, mapped from target carbon emission levels. Note that $\mathcal{J}$ and $\mathcal{R}$ satisfy Assumptions 1 and 2 respectively. Thus each user has a utility of the form (3).

In order to easily obtain the Nash equilibria of this game, we use the parametric map,

$$\mathcal{F}_i^{V_i,\pi}(x) = \mathbb{E}_{\tilde{\mathbf{q}}}[\xi V_i'(\mathcal{R}_i(x,\xi))]. \tag{12}$$

Lemma 5. *The Nash equilibrium for $\mathcal{G}_{PT}$ is $\mathbf{x}^* = (x_1^*, ..., x_N^*)$ where x_i^* is the fixed point of the decreasing function*

$$f_i(x) = d + \frac{1}{2a_i}\left(\mathcal{F}_i^{V_i,\pi}(x) - p_i\right). \tag{13}$$

Proof. The proof follows by taking the partial derivatives of $J_i^{V_i,\pi}$ and equating to zero. The derivative is equal to zero at x_i which gives the fixed-point equation

$$x_i = d + \frac{1}{2a_i}\left(\mathcal{F}_i^{V_i,\pi}(x_i) - p_i\right). \tag{14}$$

Since V_i is concave, $\mathcal{F}_i^{V_i,\pi}(x)$ is decreasing in x; so is $f_i(x)$.

Example: We plot the equilibria for the PT function given by

$$V_i(x) = \begin{cases} \log(1+x), & x \geq 0, \\ x, & x < 0. \end{cases} \tag{15}$$

The parameters are $a_i = 0.5$, $d = d_i = 1$, $\xi_1 = 1$, $\xi_2 = 2$, $\mathbf{q} = \{0.5, 0.5\}$, and $\pi(x) = x$. The solution of (14) is given by the intersection of the functions $f(x) = x$ and $f_i(x)$ given by (13). This is illustrated n Fig. (2a) at $p_i = 1.6$.

Non Concave PT Functions: While Lemma 5 identifies Nash equilibria for concave V_i, we can use the fixed point Eq. (14) to identify stationary points even when V_i is non concave. Whether these points are Nash equilibria or not, will depend on second order derivative behaviour. However, these fixed points will be the only candidates for Nash equilibria.

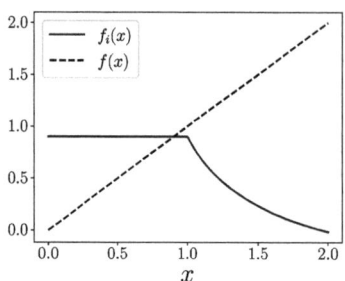 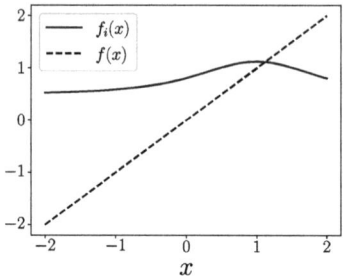

(a) Linear-logarithmic V_i with $p_i = 1.6$. (b) Logistic V_i with $p_i = 0.5$.

Fig. 2. Intersection of $f_i(x)$ and $f(x)$ for different V_i.

Example: We plot the equilibria for the logistic PT function, given by

$$V_i(x) = \frac{1}{1 + \exp(-x)} - 0.5.$$

The parameters are $a_i = 0.5$, $d = d_i = 1$, $\xi_1 = 1$, $\xi_2 = 2$, $\mathbf{q} = \{0.5, 0.5\}$, and $\pi(x) = x$. The solution of (14) is given by the intersection of the functions $f(x) = x$ and $f_i(x)$ given by (13). This is illustrated n Fig. (2b) at $p_i = 0.5$. The function $f_i(x)$ is neither increasing nor decreasing, but varies slow enough to allow only a unique intersection with $f(x)$. The second derivative of J_i at this point is negative; hence it is a local Nash equilibrium[2].

For concave V_i we also have the following parametric monotonicity of the Nash equilibrium.

Lemma 6. Let $\mathbf{x}^*(\mathbf{p})$ be the Nash equilibrium of $\mathcal{G}_{PT}$ at $\mathbf{p}$, with the strategy of agent i being $x_i^*(p_i)$. Then, $q < p_i \implies x_i^*(q) \geq x_i^*(p_i)$.

Proof. Suppose otherwise, that $x_i^*(q) < x^*(p_i)$. From (14), we see that

$$x_i^*(p_i) - x_i^*(q) = \frac{1}{2a_i}(\mathcal{F}_i^{V_i, \pi}(x_i^*(p_i)) - \mathcal{F}_i^{V_i, \pi}(x_i^*(q))) + \frac{1}{2a_i}(q - p_i).$$

Since $q < p_i$ and $\mathcal{F}_i^{V_i, \pi}(x)$ is decreasing in x, the RHS of the above equation is strictly negative, whereas the RHS is positive. Thus we have a contradiction.

4.1 Price Change from EUT to PT

The change from $\mathcal{G}_{EUT}$ to $\mathcal{G}_{PT}$ indicates a change in agent perception, a movement from a rational to a bounded rational framework. Let **x** be any equilibrium,

---

[2] A point which satisfies the maximization condition of (4) in a ball around it is called a local Nash equilibrium. In the case of utility functions that do not have global maximization properties such as concavity, such an equilibrium is a practical solution to look for.

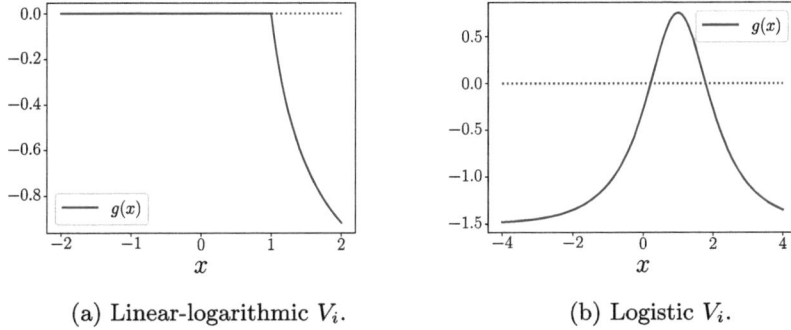

(a) Linear-logarithmic V_i. (b) Logistic V_i.

Fig. 3. Difference between EUT and PT prices for different V_i.

and let **p** and $\hat{\mathbf{p}}$ be the prices for achieving these equilibria in $\mathcal{G}_{EUT}$ and $\mathcal{G}_{PT}$ respectively. We see using (14) that

$$g(x_i) \triangleq p_i - \hat{p}_i = \mathcal{F}_i^{V_i,\pi}(x_i) - \mathbb{E}_{\mathbf{q}}[\xi]. \tag{16}$$

Example: We plot $g(x)$ in Fig. 3a for linear-logarithmic (15) V_i. Other parameters are the same as in previous examples. We see that for equilibria that are below $d_i = 1$, the same equilibrium as in $\mathcal{G}_{EUT}$ can be obtained in $\mathcal{G}_{PT}$ at the same price. However, for x_i above d_i, the corresponding price in $\mathcal{G}_{PT}$ is required to be higher. However, if one plots $g(x)$ for logistic $V_i(x) = \frac{1}{1+\exp(-6x)}$, we see (Fig. 3b) that $g(x)$ becomes positive for certain values of x.

4.2 Steering the PT Game to a Desired Point

Algorithm 1 outlines the process of steering the equilibrium in a desired direction, the target value of $\mathcal{J}(\mathbf{x})$ being α. The coordinator updates the current tariff $\mathbf{p}(n)$ using gradient descent on the function $(\mathcal{J}(\mathbf{x}) - \alpha)^2$. From the current $\mathbf{x}(n)$ it moves in a direction of descent of $(\mathcal{J}(\mathbf{x}) - \alpha)^2$. The game is steered such that it reaches an equilibrium $\mathbf{x}^*$ such that $\mathcal{J}(\mathbf{x}^*) = \alpha$. In Fig. 4, we show an example of convergence of $\mathcal{J}(\mathbf{x})$, 3 agents of whom the first has $V_i(x)$ given by (15), and the other two have $V_i(x) = x$. The parameters are $a_1 = 0.1, a_2 = 0.2, a_3 = 0.7, d = 1, \xi_1 = 1, \xi_2 = 2, \mathbf{q} = \{0.5, 0.5\},, \pi(x) = x$ and initializations $p_1 = 1, p_2 = 0.1, p_3 = 0.5$ with step size taking the constant value $\Delta(n) = 0.1$. While the theory requires a decreasing step size, we observe numerically that constant step size works well, although it requires fine tuning. Also note that $(\mathcal{J}(\mathbf{x}) - \alpha)^2$ is not a convex function. However, it has multiple global minima with identical values, and its cross section has a "W" shape, which would allow it to behave like a convex function for gradient descent, and possibly satisfy the convergence requirements of Lemma 4. The target $\alpha = -5$ is achieved to within 2% by about 250 iterations. The rate of convergence of the algorithm can be further improved by using more sophisticated gradient techniques.

Algorithm 1. Steering algorithm for game coordinator

1: $\mathbf{p}(1), \delta > 0, \Delta(n) > 0, \alpha, \hat{N}, n = 1$ ▷ Initialization
2: **while** $n \leq \hat{N}$ **do**
3: $\quad \mathbf{p} \leftarrow \mathbf{p}(n)$
4: $\quad x_i \leftarrow \text{proj}_{\mathcal{X}_i}\left[x_i + \Delta(n)\frac{\partial J_i^{V_i,\pi}(x_i,\mathbf{x}_{-i},\mathbf{p}(n))}{\partial x_i}\right]$ ▷ Update strategy at each agent i
5: $\quad D_i = \frac{\partial}{\partial x_i}(\mathcal{J}(\mathbf{x}) - \alpha)^2$ ▷ Computed by coordinator
6: $\quad p_i \leftarrow p_i + \Delta(n) D_i$ ▷ Update incentive for each agent i
7: $\quad \mathbf{p}(n+1) \leftarrow \text{proj}_{\mathcal{P}}[\mathbf{p}]$
8: $\quad n \leftarrow n + 1$
9: **end while**

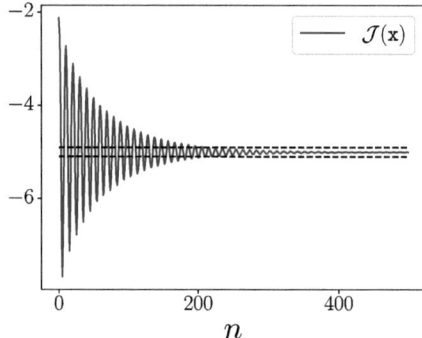

Fig. 4. Convergence of $\mathcal{J}(\mathbf{x})$ in Algorithm 1. The dashed lines indicate a 2% margin around the target.

5 Conclusion

An algorithm that uses incentives to steer a finite set of agents, combining collective and individual interests, to a collective target, was presented. Using these incentives, we can obtain targeted collective behaviour, even if the agents do not behave rationally. We conclude that when the irrational perspective affects only a component of their utility, agents can be made to reach a collective or even "rational" goal, by providing the right incentive.

Acknowledgments. This work was supported by Défi Inria-EDF.

Disclosure of Interests. The authors have no competing interests to declare that are relevant to the content of this article.

References

1. Aumann, R.J.: Rationality and bounded rationality. Games Econom. Behav. **21**(1–2), 2–14 (1997)
2. Bertsekas, D.P., Tsitsiklis, J.N.: Gradient convergence in gradient methods with errors. SIAM J. Optim. **10**(3), 627–642 (2000)
3. Ermoliev, Y.M., Flaam, S.: Repeated play of potential games. Cybern. Syst. Anal. **38**, 355–367 (2002)
4. Fochesato, M., Pokou, F., Le Cadre, H., Lygeros, J.: Noncooperative games with prospect theoretic preferences. IEEE Control Systems Letters (2025)
5. Gan, L., Hu, Y., Chen, X., Li, G., Yu, K.: Application and outlook of prospect theory applied to bounded rational power system economic decisions. IEEE Trans. Ind. Appl. **58**(3), 3227–3237 (2022)
6. Grammatico, S.: Dynamic control of agents playing aggregative games with coupling constraints. IEEE Trans. Autom. Control **62**(9), 4537–4548 (2017)
7. Holmes, R.M., Jr., Bromiley, P., Devers, C.E., Holcomb, T.R., McGuire, J.B.: Management theory applications of prospect theory: accomplishments, challenges, and opportunities. J. Manag. **37**(4), 1069–1107 (2011)
8. Jallais, S., Pradier, P.C.: The Allais paradox and its immediate consequences for expected utility theory. Routledge New York (2005)
9. Jensen, M.K.: Aggregative games and best-reply potentials. Econ. Theor. **43**(1), 45–66 (2010)
10. Kahneman, D., Tversky, A.: Prospect theory: an analysis of decision under risk. Econometrica **47**(2), 363–391 (1979)
11. Keskin, K.: Equilibrium notions for agents with cumulative prospect theory preferences. Decis. Anal. **13**(3), 192–208 (2016)
12. Krishnan K. S., A., Le Cadre, H., Bušić, A.: How Irrationality Shapes Nash Equilibria: A Prospect-Theoretic Perspective (2025). https://hal.science/hal-05036781, preprint
13. Merrick, J.R., Leclerc, P.: Modeling adversaries in counterterrorism decisions using prospect theory. Risk Anal. **36**(4), 681–693 (2016)
14. Metzger, L.P., Rieger, M.O.: Non-cooperative games with prospect theory players and dominated strategies. Games Econom. Behav. **115**, 396–409 (2019)
15. Monderer, D., Shapley, L.S.: Potential games. Games Econ. Behav. **14**(1), 124–143 (1996)
16. Rosen, J.B.: Existence and uniqueness of equilibrium points for concave n-person games. Econometrica: J. Econometric Soc. 520–534 (1965)
17. Shalev, J.: Loss aversion equilibrium. Internat. J. Game Theory **29**, 269–287 (2000)
18. Stott, H.P.: Cumulative prospect theory's functional menagerie. J. Risk Uncertain. **32**, 101–130 (2006)
19. Vahid-Pakdel, M., Ghaemi, S., Mohammadi-Ivatloo, B., Salehi, J., Siano, P.: Modeling noncooperative game of Gencos' participation in electricity markets with prospect theory. IEEE Trans. Industr. Inf. **15**(10), 5489–5496 (2019)
20. Von Neumann, J., Morgenstern, O.: Theory of games and economic behavior. Princeton University Press (1944)

Energy-Efficient Optimization of Cooperative Spectrum Sensing Algorithms in Multi-RAT Cognitive Networks

Farzam Nosrati[1], Antonio Scarvaglieri[2,3], Mariana Falco[1(✉)], Fabio Busacca[2,3], Daniele Croce[1,2], and Sergio Palazzo[2,3]

[1] University of Palermo, Palermo, Italy
{farzam.nosrati,mariana.falco,daniele.croce}@unipa.it
[2] CNIT – National Inter-University Consortium for Telecommunications, Sofia, Italy
antonio.scarvaglieri@phd.unict.it,
{fabio.busacca,sergio.palazzo}@unict.it
[3] University of Catania, Catania, Italy

Abstract. The growing demand for wireless services necessitates efficient spectrum use, especially in 5G, NextG, and Internet of Things (IoT) network scenarios. Cognitive Radio (CR) systems can be employed to detect underutilized bands, allowing Secondary Users (SUs) to opportunistically access these bands, thus improving spectrum efficiency. Cooperative Spectrum Sensing (CSS) is essential for reliable detection of Primary Users (PUs), but full SU participation is energy-intensive. This article presents an energy-efficient sensing optimization algorithm that dynamically selects an optimal SU subset by iteratively assessing marginal gains in detection probability. SUs are ranked by Signal-to-Noise Ratio (SNR), and an iterative loop with a convergence threshold determines the final sensing set. Simulations show the algorithm significantly reduces the number of active SUs, still obtaining excellent detection performance.

Keywords: Cooperative Spectrum Sensing · Cognitive Radio Networks · 5G · NextG · IoT · Energy Efficiency · Iterative Optimization · Spectrum Management

1 Introduction

The rapid advancement of technologies such as autonomous systems, smart manufacturing, water management, and smart agriculture is reshaping communication systems. These applications demand intelligent mechanisms to manage and interact with heterogeneous RF environments, especially in multi-RAT (Radio Access Technology) scenarios introduced by 5G and NextG and IoT networks [1,2]. While Cognitive Radio (CR) systems enable RF-aware adaptation, interference management remains a major challenge due to the inefficiencies of static spectrum allocation [3]. CRs rely on spectrum sensing to detect unoccupied frequency bands, or spectrum holes, thereby improving spectrum efficiency [4]. This allows Secondary Users (SUs) to identify spectrum opportunities and dynamically switch channels, reducing interference with Primary Users

(PUs). Hence, accurate and efficient detection is crucial for optimal performance in Cognitive Radio Networks (CRNs).

Energy detection, being simple and cost-effective [5], remains a widely used sensing method, but it suffers from limitations such as long sensing times, poor performance under low-SNR conditions, and inability to distinguish PU from SU transmissions [6]. These drawbacks worsen under noise uncertainty, multipath fading, and shadowing. Cooperative Spectrum Sensing (CSS) mitigates these challenges by aggregating sensing reports from multiple SUs, exploiting spatial diversity to improve detection reliability [6,7]. However, the performance of CSS depends strongly on the cooperation strategy, sensing method, reporting scheme, data fusion rule, and—critically—which SUs are selected to participate in the sensing process. Involving all SUs enhances detection accuracy but also increases sensing and reporting overhead, which raises energy consumption and reduces scalability. This has motivated a large body of work on node selection strategies in CSS. Early approaches employed SNR-based selection to improve robustness in noisy environments [5]. Threshold-based schemes allowed only high-confidence transmissions, thereby reducing unnecessary messages. Authors in [8] jointly optimize sensor selection and energy detection thresholds through a convex optimization problem. Metaheuristic optimization methods have also been employed to achieve target detection performance while balancing sensing responsibilities across nodes to prolong network lifetime [9,10]. Cluster-based and game-theoretic methods have been explored as well; for example, authors in [11] proposed a node selection scheme that considers energy, distance, and historical detection accuracy for robust decision fusion. On the other hand, authors in [12] modeled CSS as a coalition formation game in which SUs form minimal winning coalitions to achieve detection targets with reduced overhead. Oksanen *et al.* [13] applied reinforcement learning to optimize sensing policy, reducing the number of active nodes while maintaining detection reliability. Despite these advances, many CSS schemes remain static or computationally intensive, limiting their applicability in dynamic, resource-constrained environments. To address this, we propose a dynamic, iterative algorithm that adaptively selects an optimal sensing subset, maximizing detection probability while minimizing both energy consumption and computational cost.

In this work, we propose an *Energy-Efficient Iterative Sensing Optimization* algorithm, a low-complexity yet adaptive node selection method for CSS that jointly optimizes detection probability and energy use. Specifically, our contributions can be summarized as follows: (i) we introduce a heuristic, SNR-based node selection strategy that ensures high detection accuracy while eliminating unnecessary energy expenditure on nodes whose inclusion yields negligible performance gains; (ii) we provide a mathematical formulation of detection and false alarm rates for any subset of nodes, enabling optimal weighting of their contributions; and (iii) we show through simulation that our method achieves comparable or superior detection probability to full-node CSS with up to 40–60% fewer active nodes, significantly reducing sensing and reporting energy consumption.

2 Problem Statement

We consider a network, in which multiple SUs are allowed to access a licensed spectrum band, where SUs within each other's transmission range can exchange sensory data related to PU detection. The CSS scheme is illustrated in Fig. 1. Each SU is a multi-RAT nod which carries two radios: (a) **Sensing channel radio (SU-SC)**. Active during *sensing*, it samples the 10-MHz LTE uplink to detect the PU, and, if the band is free, for data transmissions to gNBs; (b) **Reporting channel radio (SU-RC)**. A 12.5-kHz FSK link reserved for collision-free control packets. It never performs wideband sensing. Thus, the SU-RC radio remains powered each slot - either in RX or sleep - for coordination, whereas the SU-SC radio is duty-cycled and wakes only for sensing and, when the cooperative decision declares the band idle, for data transmissions. Time is divided into slots of duration T, and before each data transmission, SUs must sense PU activity. However, not all nodes are reliable or effective in sensing. To address this, Sect. 3.3 formalizes our algorithm to identify and select the most valuable nodes for participation in the sensing and decision-making processes.

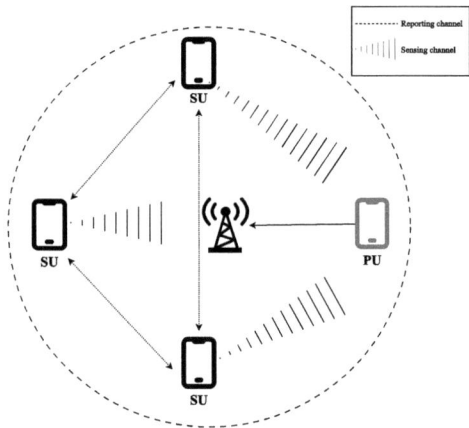

Fig. 1. An illustration of the CSS scheme, where multiple SUs detect signals from a PU.

3 Methodology

Involving all nodes in spectrum sensing can lead to high energy consumption. This work targets two main goals: maximizing detection performance while minimizing energy use. We propose an algorithm that dynamically selects an optimal subset of nodes to enhance detection in an energy-efficient CSS setup. This section presents the energy model (Sect. 3.1), sensing assumptions (Sect. 3.2), and algorithm design (Sect. 3.3).

3.1 Energy Consumption

We quantify the energy consumed by each SU during one round of cooperative sensing, aligning with our algorithm's goal of minimizing energy use while preserving accuracy. We define a per-node energy model that incorporates transmission, reception, sensing, and sleep phases. A *sensing frame* is defined as a fixed-duration interval in which all sensing tasks occur once, where its duration can be defined as:

$$T_{\text{frame}} = \underbrace{T_{\text{sen}}}_{\text{local sensing}} + \underbrace{T_{\text{rx}}}_{\text{SNR reception}} + \underbrace{T_{\text{rel}}}_{\text{SNR relaying}} + \underbrace{T_{\text{brd}}}_{\text{decision broadcast}} + \underbrace{T_{\text{slp}}}_{\text{sleep}}. \quad (1)$$

All nodes wake at frame start, perform assigned roles, then sleep until the next frame. Without loss of generality, we assume here that all SUs use identical NB-IoT transceivers [14], with hardware and timing constants shown in Table 1. Let $P_{\text{tx}}^{(\text{RC})}$ be the power emitted on the reporting channel (SU-RC), a collision-free uplink sufficient for one-hop communication (for simplicity, we assume perfect time synchronization among nodes and error-free links). Each frame includes: (a) **Local sensing** (T_{sen}): selected nodes $\mathcal{S}$ (of size k) sample PU signal and estimate instantaneous SNR γ_i; (b) **SNR reception** (T_{rx}): every node except the first in the relay chain ($i > 1$) listens to the quantised SNR forwarded by its predecessor; (c) **SNR relaying** ($T_{\text{rel}} = L(k-1)/R$): the first $k-1$ nodes in $\mathcal{S}$ relay quantised SNR hop-by-hop along chain $\mathcal{C}$, with the k-th node as destination which does not forward SNR further; (d) **Decision broadcast** ($T_{\text{brd}} = 1/R_{\text{brd}}$): the k-th node fuses received SNRs into decision $\hat{H} \in \{H_0, H_1\}$ and broadcasts it, replacing further relay transmission; and (e) **Sleep** (T_{slp}): remaining frame time in low-power mode, with radio and processor off but memory and real-time clock (RTC) active for precise wake-up, saving energy and maintaining frame alignment. The energy consumed by each SU i is as follows.

Per-Phase Energy Terms

Sensing energy. For every node $i \in \mathcal{S}$,

$$E_i^{\text{sen}} = P_{\text{RF}}^{(\text{SC})} T_{\text{sen}}. \quad (2)$$

Receive Energy. Every node listens to the global decision *except* the node that generates it (node k). In addition, every node with $i > 1$ listens once to the relayed SNR. Hence

$$E_i^{\text{rx}} = P_{\text{RF}}^{(\text{RC})} \left(\mathbf{1}_{\{i \neq k\}} \frac{1}{R_{\text{brd}}} + \mathbf{1}_{\{i > 1\}} \frac{L}{R} \right), \quad i = 1, \ldots, k. \quad (3)$$

Relaying Energy. Only nodes $1, \ldots, k-1$ transmit the quantised SNR:

$$E_i^{\text{rel}} = \left(\frac{P_{\text{tx}}^{(\text{RC})}}{\eta_{\text{PA}}} + P_{\text{cct}}^{(\text{RC})} \right) \frac{L}{R}, \quad i = 1, \ldots, k-1. \quad (4)$$

Broadcast Energy (node k).

$$E_k^{\text{brd}} = \left(\frac{P_{\text{tx}}^{(\text{RC})}}{\eta_{\text{PA}}} + P_{\text{cct}}^{(\text{RC})} \right) \frac{1}{R_{\text{brd}}}. \quad (5)$$

Table 1. Radio constants for the PU and the two radios embedded in each SU: NB-IoT (SU-NB) and the narrowband reporting radio (SU-RC).

Symbol	Description	Unit	PU	SU-SC	SU-RC
$P_{\text{TX,max}}$	Max. radiated power	dBm	+23	+23	+10
P_{TX}	Radiated power	dBm	−32	—	—
P_{RF}	RX front-end + BB power	mW	400	205	12
P_{cct}	BB/PLL power in TX mode	mW	120	65	18
P_{sleep}	Stand-by/PSM power	µW	7000	3.3	30
η_{PA}	PA drain efficiency	—	0.33	0.38	0.46
B	Channel bandwidth	kHz	10000	180	12.5
N_F	Noise figure	dB	6	4.5	6.5
P_{sens}	Reference sensitivity	dBm	−101	−114	−120

Sleep Energy. A node sleeps until it must receive the next message and again after it has transmitted, so the interval depends on its position in the relay chain:

$$T_{\text{slp},i} = T_{\text{frame}} - T_{\text{sen}} - \mathbb{1}_{\{i>1\}}T_{\text{rx}} - T_{\text{rel}} - (k-i)(T_{\text{rx}} + T_{\text{rel}}) - \mathbb{1}_{\{i=k\}}T_{\text{brd}} \quad (6)$$

where $(k-i)(T_{\text{rx}} + T_{\text{rel}})$ represents the duration during which node i can remain idle (in sleep mode) while downstream nodes complete their receive–relay operations; for node 1, this interval constitutes its entire sleep phase.

$$E_i^{\text{slp}} = P_{\text{sleep}}^{(\text{RC})} T_{\text{slp},i}. \quad (7)$$

Total per-Node and Network Energy. The total energy for node i is

$$E_i = \begin{cases} E_i^{\text{sen}} + E_i^{\text{rel}} + E_i^{\text{rx}} + E_i^{\text{slp}}, & i \in \mathcal{S}, i < k \\ E_k^{\text{sen}} + E_k^{\text{rel}} + E_k^{\text{brd}} + E_k^{\text{slp}}, & i = k \\ E_i^{\text{rx}} + E_i^{\text{slp}}, & i \notin \mathcal{S} \end{cases} \quad (8)$$

and the energy over the whole network is then

$$E_\Sigma(k) = \sum_{i=1}^{N} E_i. \quad (9)$$

Since E_i is constant for nodes not in $\mathcal{S}$, minimising $E_\Sigma(k)$ reduces to selecting the smallest k that meets the detection constraints in Sect. 3.3.

Algorithm 1. Energy-Efficient Iterative Sensing Optimization

1: **Input:**
2:     Network of N nodes
3:     False alarm threshold ϵ
4:     Convergence threshold λ_{th}
5: **Phase I: Information Collection**
6: **for** $n = 1$ to N **do**
7:     Measure $\text{SNR}[n]$
8: **end for**
9: Sort nodes by $\text{SNR}[n]$ in descending order $\rightarrow$ `SortedNodes`
10: **Phase II: Iterative Sensing Optimization**
11: Initialize $\omega \leftarrow 1, P_d^{(0)} \leftarrow 0, \lambda^{(1)} \leftarrow 1, k \leftarrow 1, i \leftarrow 1$
12: **while** $\lambda^{(k)} > \lambda_{\text{th}}$ **and** $i \leq N$ **do**
13:     Select top i nodes from `SortedNodes` $\rightarrow$ `NodeSet`$_i$
14:     Optimize detection probability $P_d^{(k)}$ over `NodeSet`$_i$ subject to false alarm probability $\leq \epsilon$
15:     $\lambda^{(k+1)} \leftarrow |P_d^{(k)} - P_d^{(k-1)}|$
16:     $P_d^{(k-1)} \leftarrow P_d^{(k)}$
17:     $k \leftarrow k+1, i \leftarrow i+1$
18: **end while**
19: **Output:** Optimized node set and final detection probability $P_d^{(k)}$

Boot-Strapping Phase. In the initial frame ($r = 0$) all N SUs are active and forward their own SNR, so $k = N$. The one-off cost becomes

$$E_\Sigma^{(0)} = NP_{\text{RF}}^{(\text{SC})} T_{\text{sen}} + NP_{\text{sleep}}^{(\text{RC})} T_{\text{slp}}$$
$$+ N\left(\frac{P_{\text{tx}}^{(\text{RC})}}{\eta_{\text{PA}}} + P_{\text{cct}}^{(\text{RC})}\right)\frac{L}{R} + (N-1)P_{\text{RF}}^{(\text{RC})}\frac{L}{R}$$
$$+ NP_{\text{RF}}^{(\text{RC})}\frac{1}{R_{\text{brd}}} + \left(\frac{P_{\text{tx}}^{(\text{RC})}}{\eta_{\text{PA}}} + P_{\text{cct}}^{(\text{RC})}\right)\frac{1}{R_{\text{brd}}}. \quad (10)$$

As the number of iterations grows, this start-up cost is amortised: over M rounds the mean per-frame energy is $\left(E_\Sigma^{(0)} + \sum_{r=1}^{M-1} E_\Sigma(k_r)\right)/M$, which quickly converges to the steady-state value governed by $E_\Sigma(k)$.

3.2 Mathematical Model for Cooperative Spectrum Sensing

In spectrum access, reliably detecting the presence of a PU is critical to avoid interference while enabling efficient secondary spectrum usage. In realistic environments, individual SUs often experience deep fading, shadowing, or low SNRs, which degrade detection reliability. CSS addresses this limitation by allowing multiple SUs to share their sensing information, thereby exploiting spatial diversity to mitigate fading and shadowing effects and improve overall detection reliability. The cooperative sensing process starts with each SU performing a local binary hypothesis test to detect the presence H_1 or absence H_0 of a PU signal, modeled as:

$$x_i(j) = \begin{cases} \nu_i(j), & \text{if } H_0 \text{ holds} \\ h_i s(j) + \nu_i(j), & \text{if } H_1 \text{ holds} \end{cases} \quad (11)$$

where $x_i(j)$ is the signal at the i-th SU, $s(j)$ is the PU signal, and $\nu_i(j)$ is zero-mean additive white Gaussian noise (AWGN) experienced by the i-th SU. The channel gain h_i reflects fading and shadowing, modeled as $h_i(j) = \sqrt{\beta \left(\frac{d_0}{d_i}\right)^\alpha 10^{-\frac{\psi_i}{10}}}$; where β is a system constant, α the path-loss exponent, d_i the SU-PU distance, d_0 a reference distance, and ψ_i is shadow fading with variance σ_i^2. We adopt energy detection for CSS, where each SU computes and energy statistic forming the set $\mathbf{y} = \{T_1(\mathbf{x}), T_2(\mathbf{x}), \ldots, T_i(\mathbf{x})\}$, with:

$$T_i(\mathbf{x}) = \sum_{j=0}^{J-1} |x(j)|^2 \underset{H_0}{\overset{H_1}{\gtrless}} \gamma_i, \quad (12)$$

and γ_i as the detection threshold. For $J \geq 20$, the energy statistic approximates a Gaussian distribution as expected from the central limit theorem [15], i.e. $\mathcal{N}_0(\mu_0, \Sigma_0)$ under H_0, and $\mathcal{N}_1(\mu_1, \Sigma_1)$ under H_1, where μ_l and Σ_l (for $l \in \{0, 1\}$) are the respective mean and variance. A global decision is defined using a weight vector ω, prioritizing SUs with higher SNRs [16]. Performance is evaluated with $P_d \triangleq \Pr(\widehat{H} = H_1 \mid H_1)$ and false alarm probability $P_f \triangleq \Pr(\widehat{H} = H_1 \mid H_0)$. With ϵ bounding the acceptable false alarm rate, sensing is optimized by:

$$\max_\omega \ P_d \quad \text{subject to} \quad P_f \leq \epsilon, \quad (13)$$

3.3 The Optimization Algorithm: Goal and Characteristics

Efficient CSS in energy-constrained networks requires balancing detection accuracy with energy consumption, as activating more nodes improves the probability of detection P_d but also increases energy use. To address this, we introduce an *Energy-Efficient Iterative Sensing Optimization* method, which operates in two phases. In the *Information Collection* phase, nodes estimate their local SNR via energy detection, and share these values with a fusion center or among neighbors to reduce redundancy. In the *Sequential Sensing Optimization* phase, nodes are ranked in descending SNR order and added iteratively; at each step k, the subset of size k is used to maximize $P_d^{(k)}$ subject to the false alarm constraint $P_f \leq \epsilon$. The process stops when the marginal gain in detection performance falls below a predefined threshold λ_{th}, yielding a heuristic, threshold-based selection strategy that avoids the computational complexity of exhaustive search. Finally, detection and false alarm rates for the selected subset of size k are given by:

$$P_k^f \triangleq \Pr(\widehat{H} = H_1 \mid H_0) = Q\left(\frac{\gamma_k - \mu_{0k}^\top \omega_k}{\sqrt{\omega_k^\top \Sigma_{0k}\, \omega_k}}\right),$$
$$P_k^d \triangleq \Pr(\widehat{H} = H_1 \mid H_1) = Q\left(\frac{\gamma_k - \mu_{1k}^\top \omega_k}{\sqrt{\omega_k^\top \Sigma_{1k}\, \omega_k}}\right) \tag{14}$$

where $Q(\cdot)$ is the Gaussian Q-function, $\mu_{0k} = J\left[\sigma_1^2, \sigma_2^2, \ldots, \sigma_k^2\right]^\top$ and $\mu_{1k} = \left[(J+\eta_1)\sigma_1^2, (J+\eta_2)\sigma_2^2, \ldots, (J+\eta_k)\sigma_k^2\right]^\top$, and $\Sigma_{0k} = 2J \cdot \text{diag}[\sigma_1^4, \sigma_2^4, \ldots, \sigma_k^4]$, $\Sigma_{1k} = 2J \cdot \text{diag}[2\sigma_1^4(J+2\eta_1), 2\sigma_2^4(J+2\eta_2), \ldots, 2\sigma_i^4(J+2\eta_k)]$ represent the mean vectors and covariance matrices up to k nodes under hypotheses H_0 and H_1. ω_k is the weighting vector over the selected nodes, and η_i is the SNR at the i-th SU node calculated as: $\eta_i = |h_i|^2/\sigma_i^2 \cdot \sum_{j=0}^{J-1} |s_j|^2$. At each iteration, when a new sensor is added, the optimization $\max_{\omega_k} P_d^k$ is performed to obtain the optimal weight for node k. The improvement between iteration is $\lambda^k = |P_d^k - P_d^{k-1}|$, and the process stops when $\lambda^k \leq \lambda_{\text{th}}$. The final result is an optimized sensing node set, balancing detection performance and energy use, which can be formally stated as

$$\min_{k=\{1,\ldots,N\}} E_\Sigma(k) \quad \text{subject to} \quad \lambda^k \leq \lambda_{\text{th}}. \tag{15}$$

The proposed SNR-based optimization ensures rapid convergence of the algorithm: as empirically shown in Fig. 2, the marginal gain in detection probability, λ^k, diminishes sharply when nodes are added in descending order of their SNR. Consequently, the stop criterion $\lambda^k \leq \lambda_{th}$ is typically met after only a small fraction of the total nodes have been included, ensuring a low number of iterations. The computational complexity of Algorithm 1 comprises four main components: sorting, sequential node selection, broadcasting, and convex subproblem optimization. Sorting the SUs by SNR requires

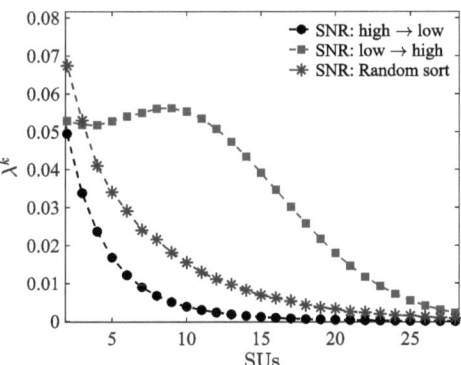

Fig. 2. Detection improvement rate λ^k versus the number of participating secondary users (SUs), for a network of 30 SUs. The red squares represent SUs sorted in ascending SNR order, black circles represent descending SNR order, and blue stars indicate random node selection (averaged over a large number of simulations). (Color figure online)

$O(N \log N)$ time. Sequential selection proceeds for at most $k_{\max}$ rounds (adding one node per round), resulting in $O(k_{\max})$ round complexity and $O(Nk_{\max})$ message complexity due to fusion-center broadcasts. For the subset of size k at each round, the optimal weight vector is obtained by solving a convex subproblem; with warm-started interior-point updates, this step requires $O(k_{\max}^2)$ time across all rounds.

4 Numerical Results

To validate empirically the design choice of sorting nodes by descending SNR in Algorithm 1, we demonstrate how this ordering enables effective identification of a subset of nodes that maximizes detection performance with minimal energy expenditure. We simulate a network of 30 secondary user nodes, each receiving signals with SNR values uniformly distributed between -22 dB and 5.5 dB. The channel model incorporates free-space path loss ($\alpha = 2$) and random shadowing of ± 5 dB. We assume the PU radiates an equivalent isotopically radiated power of $P_{\text{TX}}^{(\text{PU})} = -32$ dBm on the LTE Band 20 up-link ($f_c = 806$ MHz, $B = 10$ MHz). This setting reflects the low-power indoor small-cell class in 3GPP TS 36.101 [14], used in all simulations. We assess the incremental gain in detection probability, defined as $\lambda^k = |P_d^{(k)} - P_d^{(k-1)}|$, for consecutive SUs at a fixed false alarm probability. As shown in Fig. 2, sorting SNR values in descending order yields a rapidly decreasing λ^k, enabling the definition of a practical threshold λ_{th} for selecting the most impactful nodes. Ascending order does not exhibit a consistent decay in λ^k, complicating cutoff selection. Random ordering shows a generally decreasing trend but with more variability.

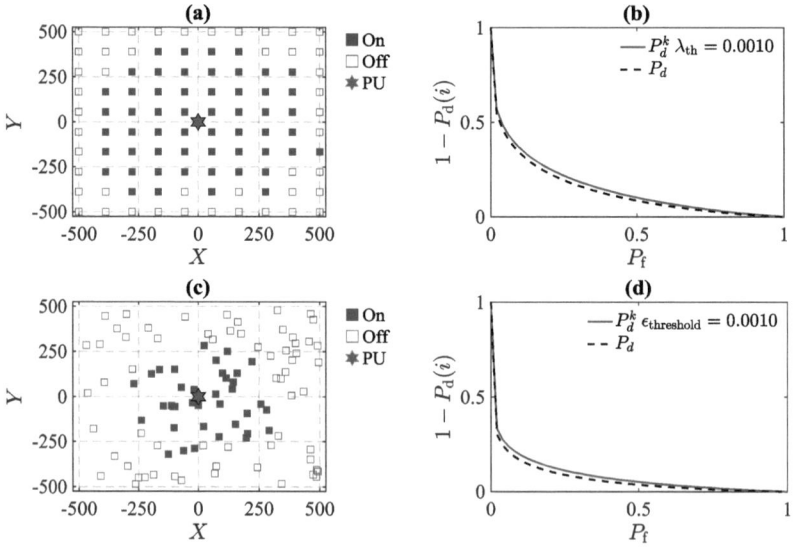

Fig. 3. Uniform (a) and random (c) placement of 100 SUs in a 1 km × 1 km area, with a PU at the center. Miss detection probability versus false alarm probability in case of uniform (b) and random (d) placement, comparing threshold $\lambda_{\text{th}} = 0.001$ against all-node participation.

To evaluate the approach under realistic conditions, we simulate a 1 km^2 area with 100 SUs placed either uniformly or randomly, and one PU at the center of the area, as shown in Fig. 3(a) and Fig. 3(c) respectively. We minimize the miss detection probability under a fixed false alarm constraint, with and without node selection. Using $\lambda_{th} = 0.001$, Fig. 3(b) and (d) show that 50% (uniform) and 69% (random) SUs can be excluded without significant performance loss. This yields energy savings of about 47% and 64%, respectively, compared to the baseline where all nodes participate.

5 Conclusions

This article proposed an Energy-Based Iterative Sensing Optimization algorithm to improve CSS efficiency in multi-RAT Cognitive Networks. By iteratively selecting a subset of SUs based on their SNR and optimizing detection probability, the method reduces energy consumption while maintaining performance. Nodes are ranked and added incrementally until the marginal detection gain falls below a predefined threshold. Simulations demonstrate that the algorithm significantly reduces the number of active SUs, without sacrificing accuracy, yielding energy savings of 47% (uniform grid) and 64% (random layout) in single PU scenarios. Future directions include addressing mobility-induced dynamics, exploring game-theoretic models for decentralized decision-making, and evaluating robustness under imperfect reporting and heterogeneous SU capabilities.

Acknowledgments. This work was supported by the European Union under the Italian National Recovery and Resilience Plan (NRRP) of NextGenerationEU: partnership on "Telecommunications of the Future" (PE00000001 - program "RESTART"), S2 SUPER – Programmable Networks, Cascade project PRISM - CUP: C79J24000190004; and project PNRR M4 - C2 - investment 1.1: Projects of Significant National Interest (PRIN) - PRIN 2022 PNRR code P2022WA578 "BISS: Beyond-5G Infrastructure for Spectrum Sensing", CUP B53D23024110001 and PRIN 2022 code 2022FYCNPT "IoTSensE: IoT-based Sensing Extension", CUP B53D23002610006.

References

1. Akram, T., Esemann, T., Hellbrück, H.: Cooperative spectrum sensing protocols and evaluation with IEEE 802.15. 4 devices. Phys. Commun. **19**, 93–105 (2016)
2. Pagano, A., et al.: A survey on massive IoT for water distribution systems: challenges, simulation tools, and guidelines for large-scale deployment. Ad Hoc Netw. **168**, 103714 (2025)
3. Garlisi, D., Pagano, A., Giuliano, F., Croce, D., Tinnirello, I.: Interference analysis of LoRaWAN and Sigfox in large-scale urban IoT networks. IEEE Access **13**, 44836–44848 (2025)
4. Mitola, J.: Cognitive radio for flexible mobile multimedia communications. In: 1999 IEEE International Workshop on Mobile Multimedia Communications (MoMuC'99). IEEE (1999)
5. Atapattu, S., Tellambura, C., Jiang, H.: Energy detection for spectrum sensing in cognitive radio, volume 6. Springer (2014)
6. Akyildiz, I.F., Lo, B.F., Balakrishnan, R.: Cooperative spectrum sensing in cognitive radio networks: a survey. Phys. Commun. **4**(1), 40–62 (2011)

7. Nosrati, F., Gelaw, E., Corallo, R., Schilleci, S., Vicario, A., Croce, D.: Cooperative spectrum sensing for beyond-5g networks in fading environments. In: Proceedings of the Twenty-Fifth International Symposium on Theory, Algorithmic Foundations, and Protocol Design for Mobile Networks and Mobile Computing, MobiHoc '24, pp. 446–451, New York, NY, USA (2024). Association for Computing Machinery
8. Ebrahimzadeh, A., Najimi, M., Fallahi, A.: Sensor selection and optimal energy detection threshold for efficient cooperative spectrum sensing. IEEE Trans. Veh. Technol. **64**(4), 1565–1577 (2015)
9. Shrestha, A., Bajracharya, C., Kim, S.W.: An energy efficient fair node selection for cooperative in-band and out-of-band spectrum sensing. Comput. Commun. **119**, 83–93 (2018)
10. He, J., Jiang, H., Wang, C., Wen, M.: Evolutionary search for energy-efficient distributed cooperative spectrum sensing. Ad Hoc Netw. **102**, 102100 (2020)
11. Jin, Z., Qiao, Y.: A novel node selection scheme for energy-efficient cooperative spectrum sensing using D-S theory. Wireless Netw. **26**, 269–281 (2020)
12. Saad, W., Han, Z., Debbah, M., Hjorungnes, A., Basar, T.: Coalition formation games for collaborative spectrum sensing. In: IEEE International Conference on Communications (ICC), pp. 1–5. IEEE (2010)
13. Oksanen, J., Cabric, D.: Reinforcement learning based sensing policy optimization for energy efficient cognitive radio networks. IEEE Global Telecommun. Conf. (GLOBECOM) 1–6 (2011)
14. 3rd Generation Partnership Project (3GPP). E-UTRA; narrowband IoT (NB-IoT); UE radio transmission and reception (2024)
15. Gnedenko, B.V., Kolmogorov, A.N.: Limit distributions for sums of independent random variables, volume 2420. Addison-Wesley (1968)
16. Quan, Z., Cui, S., Sayed, A.H.: Optimal linear cooperation for spectrum sensing in cognitive radio networks. IEEE J. Sel. Top. Sig. Process. **2**(1), 28–40 (2008)

Coordinated Attack Planning in Probabilistic Attack Graphs within a Sensor-Allocated Network

Romaric Mofouet[1(✉)], Haoxiang Ma[2], Jie Fu[2], Charles Kamhoua[3], Gabriel Deugoue[1], and Arnold Kouam[4]

[1] University of Dschang, Dschang, Cameroon
mofouet@gmail.com
[2] Department of Electrical and Computer Engineering, University of Florida, Gainesville , USA
hma2@ufl.edu, jfu2@wpi.edu
[3] DEVCOM Army Research Laboratory, Adelphi, USA
charles.a.kamhoua.civ@army.mil
[4] Avignon University, Avignon, France
arnold.kouam-kounchou@univ-avignon.fr

Abstract. Modern cyberattacks increasingly involve coordinated teams of adversaries, posing new challenges for detection and defense. This paper introduces a probabilistic framework for modeling such multi-agent attacks, combining Markov Decision Processes with an augmented Markov game that captures dynamic team composition under detection events. The model integrates state-dependent detection probabilities, implicit coordination via reward structures, and efficient value iteration for policy computation. Through a stochastic grid-world case study, we analyze the performance trade-offs between isolated and coordinated strategies. Our results reveal that while isolated agents perform better in simple, single-target scenarios, coordinated strategies significantly improve success rates in complex, multi-target operations by systematically managing exposure. This work provides a formal foundation and practical methodology for analyzing emergent attack behaviors in adversarial environments.

Keywords: Multi-agent · Coordinated attack · Markov process · Value iteration

1 Introduction

As the number of interconnected systems grows, cyberattacks have become more sophisticated, often orchestrated by teams of adversaries working together to increase their chances of success. These coordinated attacks, such as stealthy scanning, worm propagation, or DDoS campaigns, are particularly hard to stop when attackers adopt diverse behaviors and adapt to partial detection [6]. These scenarios illustrate the need for quantitative models that not only capture how coordinated agents progress in an uncertain environment, but also how they degrade as detection mechanisms progressively neutralize them. Providing such models is crucial to inform defensive resource allocation, sensor placement, and risk assessment in realistic cyber environments.

In this paper, we study how a group of attackers can strategically coordinate their actions to reach a common objective in a network equipped with a static detection mechanism. We focus on the planning problem faced by such a team: "how to maximize the chance that each agent reach a distinct goal before all agents are neutralized". Detection is probabilistic, varies across the environment, and results in an agent becoming inactive (moved to a **dummy state**) while remaining in the system. We model this problem using a Markov Decision Process (MDP) framework extended to multiple agents. Each agent's planning is modeled as a probabilistic attack graph, and the team operates in an environment where their composition may evolve as detections occur. Our key originality lies in explicitly modeling detection-driven attrition through absorbing dummy states, which allows us to capture the progressive neutralization of attackers, a feature absent from prior MDP or attack graph formulations. To capture this, we introduce an augmented Markov game that tracks the set of active agents, updating transitions accordingly. In this game, a dummy state represents neutralized attackers, and the game ends either when at least one goal is reached or all agents are neutralized.

Related work includes probabilistic attack graphs [8,10,12], game-theoretic defenses [7,13], and Markov games for security [20]. These assume static attacker teams, focus on defender policies, or lack detection-induced attrition mechanisms. Coordination models [2] omit probabilistic detection. Recent efforts have revisited attack graphs and security MDPs using stochastic optimization and learning-based methods [3,16], and our work complements these by explicitly modeling coordinated attacks under detection-driven team attrition using spatial-aware Markov games. In contrast, our contribution unifies these strands by focusing explicitly on the attackers' strategy optimization using a probabilistic, multi-agent framework that incorporates dynamic neutralization of attackers. We illustrate our model through a grid-world case study, where detection probabilities vary spatially. For instance, during a DDoS-style simulation, agents are gradually detected and neutralized, and the remaining team must adjust strategies accordingly, mirroring real-world threats where losing one attacker doesn't stop the others. Specifically, (1) we propose a formal augmented Markov game model that integrates detection-aware team attrition and state-dependent sensing. (2) We present a reward modeling scheme that distinguishes partial and full mission success, allowing fine control over coordination incentives. (3) We validate the model through grid-world simulations, revealing the trade-offs between resilience, mission completeness, and exposure risk under different coordination strategies.

2 Framework and System Description

Consider a network of agents that are associated with an underlying undirected graph $G = (\mathcal{N}, \mathcal{E})$ where $\mathcal{N} = \{1, 2, \ldots, n\}$ denotes the set of agents (n the number of agents) and $\mathcal{E}$ denotes the set of edges (how information is implicitly exchanged between agents). This team of agents aims to attack a network to gather critical resources, gain elevated privileges, or achieve other strategic objectives. However, the coordinated nature of the attack involves multiple agents with diverse behaviors, making it harder to counteract. The targeted network includes a detection mechanism capable of expelling intruders, whose locations are assumed known by the team after stealthy reconnaissance [14]. Before the detection mechanism alters the environment's transitions, each agent i ($i = 1, \ldots, n$) plans its attack as a probabilistic attack MDP $M_i = (S_i, A_i, P_i, \nu_i, \gamma, F, R_i)$, capturing environmental uncertainty. Here, S_i and A_i are the agent's local states and actions, with $s_i \in S_i$ denoting a network position and $a_i \in A_i$ representing an exploit or compromise. Being at state $s_i \in S_i$, an action $a_i \in A_i$ move the agent to s'_i with probability $P(s'_i|s_i, a_i)$. The agent's starts at position $\nu_i \in S_i$. The attack continues with probability $\gamma \in [0, 1]$, and $R_i : S_i \times A_i \to \mathbb{R}$ defines rewards based on a target set F. Each agent seeks an optimal policy $\pi_i : S_i \to \Delta(A_i)$ maximizing:

$$V^{\pi_i}(s_i) = \mathbb{E}_{\pi_i} \left[\sum_{p=0}^{+\infty} \gamma^p R_i(s_i^{(p)}, \pi_i(s_i^{(p)})) \mid s^{(0)} = s_i \right] \quad (1)$$

Unlike isolated attackers, this team can exchange information [1] and thus can infer all the team members positions in real time. In this case, coordinated agent's policies are based on global states $\boldsymbol{s} = (s_1, \ldots, s_n) \in \boldsymbol{\mathcal{S}} = S_1 \times \cdots \times S_n$, with $\pi_i : \boldsymbol{\mathcal{S}} \to \Delta(A_i)$. Since the agents are acting in the same network, we assume that:

Assumption 1. *All agents share the same environment model, i.e., $M_i = (S, A, P, \nu_i, \gamma, F, R)$, differing only in initial state ν_i.*

The defender's fixed sensor deployment is represented by a binary vector $\boldsymbol{y} \in 2^S$, where $\boldsymbol{y}(s) = 1$ indicates that a sensor is placed at state s. Accordingly, the operation of the detection mechanism is defined by the following detection function:

Definition 1 (Detection Mechanism). *Given allocation $\boldsymbol{y}$, the function $\eta^{\boldsymbol{y}} : \boldsymbol{\mathcal{S}} \times \mathcal{N} \to [0, 1]$ maps a joint state and agent index to the detection probability of agent i.*

If an agent is detected, it is removed from the system. The remaining agents define a subsystem:

Definition 2 (Subsystem). *Given agent set $X \subseteq \mathcal{N}$, define $M(X) = (S_X, A_X, P_X)$ with $S_X = \prod_{i \in X} S_i$, $A_X = \prod_{i \in X} A_i$, and $P_X(s, a, s') = \prod_{i \in X} P_i(s'_i|s_i, a_i)$.*

We assume that the defender configure sensors before the attack, so $\eta^y = \eta$ is fixed.

In many coordinated attacks, partial success (a single agent reaching a target) can offer limited value (public visibility, minor data leaks), and still stop the attack, while total strategic gain often depends on agents reaching targets in sync. This is critical in scenarios such as threshold-based crypto fraud or distributed authentication systems, where a single uncoordinated action triggers countermeasures [4,9], as well as in cyber-physical systems where only tightly coordinated stealth attacks can bypass detection and inflict significant damage [11,15]. The team objective is then:

Problem 1 (Coordinated Planning Problem). Given a multi-agent system and a detection function η, compute a strategy profile $\boldsymbol{\pi} = (\pi_1, \ldots, \pi_n)$ that maximizes the probability that all agents in the team simultaneously reach target states in the goal set F.

3 Main Approach

In the rest of the paper, we will denote by $x = |X|$ the number of elements of the set $X \subseteq \mathcal{N}$, and $\bar{x} = |\bar{X}|$, the number of elements of the comlementary of X in $\mathcal{N}$. For a tuple $\boldsymbol{s} = (s_1, s_2, \ldots, s_n)$, we use the notation $s \sqsubseteq \boldsymbol{s}$ to means that s is one of the s_i, $i \in \{1, 2, \ldots, n\}$. For a set B, we denote $\mathbb{1}_B$, the indicator function of the set B define as $\mathbb{1}_B(b) = 1$ if $b \in B$ and 0 otherwise.

3.1 Model Formulation

To keep the number of agents in the network constant while accounting for those who have been ejected, we consider these ejected players as still present but passive, meaning their actions do not influence the environment's dynamics. We introduce a new absorbing state, s^{dum}, termed the **dummy** state (an inactive absorbing state from which no actions can be taken), to represent these ejected agents, thereby distinguishing between active and passive agents. When an agent acts alone in the network, the detection function affects its local state space and transition function as follows:

Definition 3. *Given an agent i with its local attack graph $M_i = (S, A, P, \nu_i, \gamma, F, R)$, and a detection function η, the agent's attack graph become $M_i(\eta) = (S^\eta, A, P^\eta, \nu_i, \gamma, F, R^\eta)$, where: A, ν_i, γ, F are the same as in M_i,*
$$S^\eta = S \cup \{s^{dum}\}, \quad P^\eta(s'|s, a) = \begin{cases} \eta(s, 1) & \text{if } s' = s^{dum}, \\ (1 - \eta(s, 1))P(s'|s, a) & \text{otherwise, and} \end{cases}$$
$R^\eta(s, a) = (1 - 2.\mathbb{1}_{\{s^{dum}\}}(s))R(s, a).$

The definition (3) can be explained as follows: the newly created state s^{dum} is added to the state space to account for when the player is ejected from the network. With probability $\eta(s, 1)$, the agent transitions to the dummy state s^{dum}. Otherwise, the agent transitions to the next state s' with probability

$(1 - \eta(s, 1))P(s'|s, a)$. The agent receives a reward $R(s, a)$ as in M_i if the state s is not the dummy state s^{dum} and conversely, the agent is penalized samely (reward of $-R(s, a)$) if ejected.

The objective is to determine the team's optimal policy that maximizes the probability of achieving their objective. This framework accounts for all possible joint states of active agents and their corresponding joint actions. The transition dynamics are constructed based on the local transition functions of individual agents, the probability of agents remaining active at a given stage, and the likelihood of others being removed. The attack concludes either when at least one agent successfully reaches a target state or when all agents are removed from the system. Achieving several distinct targets yields substantially higher rewards under this framework compared to hitting a single target. These dynamics are formally encapsulated in the following Augmented Markov Game.

Definition 4 (Augmented Markov Game).

The switching Markov game in which the number of actors changes given the triggered detection can be constructed as $\mathcal{M} = (\mathcal{S}_M, \mathcal{A}_M, \mathcal{P}^\eta, \nu, \gamma, \mathcal{R})$, where :

- $\mathcal{S}_M = \bigcup_{X \in 2^\mathcal{N}} S_X \times \{s^{dum}_{\bar{X}}\}$, *is the state space.* S_X *represents the set of states for the subsystem with X remaining players, and $s^{dum}_{\bar{X}}$ is a vector of $\bar{x}$ dummy states. $S_\emptyset \times \{s^{dum}_{\bar{X}}\}$ denotes the joint state $\boldsymbol{s^{dum}} = (s^{dum}, s^{dum}, \ldots, s^{dum})$, where all players have been removed and are thus acting as dummies. This state is terminal in the game. If at least one agent reaches a state in F, the game transitions to a terminal state. We define $\mathcal{F} = \{(s_1, s_2, \ldots, s_n) \in \mathcal{S}_M \mid \exists i, s_i \in F\}$ as the reachability set. The complete set of terminal states is $\mathcal{S}_{term} = \mathcal{F} \cup \{\boldsymbol{s^{dum}}\}$. Since the tuple of dummy states is fixed based on the set of remaining players, we will use the notation S_X to represent this, keeping in mind that the removed players are acting passively.*
- $\mathcal{A}_M = \mathcal{A}$, *where the set of actions in the augmented MDP represents the global set of actions.*
- $\mathcal{P}^\eta : \mathcal{S}_M \times \mathcal{A}_M \times \mathcal{S}_M \to [0, 1]$ *is the transition function given the detection function η. To keep the framework manageable, we will assume that no new agents can join during the process and that once an agent is ejected into the absorbing state, they cannot leave it. Therefore, given $s_{X_1} = (s_1, \ldots, s_{x_1}) \in S_{X_1}$, $s'_{X_2} = (s'_1, \ldots, s'_{x_2}) \in S_{X_2}$, $a_{X_1} = (a_1, \ldots, a_{x_1}) \in \mathcal{A}_{X_1}$, $\eta(s_{X_1}, a_{X_1}, i)$ the probability of detecting player i, ($i \in X_1$) and that the remaining players (players in X_1) has played a joint action a_{X_1} in a joint state s_{X_1}, the probability of transiting from a framework with fewer agents (not acting dummy) to a one with more agents is 0, i.e.*

$$\mathcal{P}^\eta(s'_{X_2}|s_{X_1}, a_{X_1}) = 0 \text{ if } X_2 \not\subseteq X_1.$$

Otherwise, the probability of transiting from a framework with x_1 agents to a framework x_2 (with $x_1 \geq x_2$) is defined as

$$\mathcal{P}^\eta(s'_{X_2}|s_{X_1}, a_{X_1}) = \prod_{i \in X_1 \setminus X_2} \eta(s_{X_1}, a_{X_1}, i) \cdot \prod_{j \in X_2} \left(1 - \eta(s_{X_2}, a_{X_2}, j)\right). \quad (2)$$
$$P(s'_j|s_j, a_j) \text{ if } X_2 \subseteq X_1.$$

The Definition (2) can be justified as follows. Since the set of remaining agents is X_2, we use the joint transition of these agents and multiply it by the probability that these agents are remaining and by the probability that the agent in $X_1 \setminus X_2$ has been removed. Notice that when exactly one agent remains active in the framework, the transition function aligns precisely with the one defined in Definition 3.

- $\boldsymbol{\nu} = (\nu_1, \ldots, \nu_n) \in \underbrace{S \times S \times \cdots \times S}_{\times n}$ is the team's global initial position.
- $\gamma \in (0, 1]$ is a discount factor,
- $\mathcal{R} : \boldsymbol{S}_M \times \boldsymbol{A}_M \to \mathbb{R}$ is the joint reward function for all the agents.

The team aims to have all members simultaneously reach target states prior to any ejection, thus the reward gained strictly increase with the number of distinct goals achieved. For $\boldsymbol{s} \in \boldsymbol{S}_M$, let $\mathcal{T}_{\boldsymbol{s}} = \{s \sqsubseteq \boldsymbol{s}, s = s^{dum}\}$ the set of passive states in $\boldsymbol{s}$, and $G_{\boldsymbol{s}} = \{s \sqsubseteq \boldsymbol{s}, s \in F\}$ the set of goals achieved at the state $\boldsymbol{s}$. We then define the reward as

$$\mathcal{R}(\boldsymbol{s}, \boldsymbol{a}) = \begin{cases} 1 & if \ |G_{\boldsymbol{s}}| = n \\ r.|G_{\boldsymbol{s}}| & if \ 0 < |G_{\boldsymbol{s}}| < n \\ -\mathbb{1}_{\mathcal{T}_{\boldsymbol{s}}}(\boldsymbol{s}) & otherwise. \end{cases}$$

where $0 < r << \frac{1}{n}$ emphasizing the importance of full goal achievement. The team receives a reward of 1 if all agents reach a goal state, $r.|G(\boldsymbol{s})|$ if at least one agent reaches a goal, -1 if all agents are ejected, and 0 otherwise. The parameter r can be tuned to reflect the relative importance of partial success.

3.2 Computing Optimal Policies

One approach to tackle a multi-agent problem involves treating the resulting joint problem as an MDP, wherein the states and actions spaces are the direct product of all agents' sets of states and actions [5]. In such a context, it suffices to consider Markovian policies, as the optimal policies are always Markovian [17]. Given a joint policy $\pi : \mathcal{S} \to \mathcal{A}$, the agent's common state value function $V^\pi : \mathcal{S} \to \mathcal{R}$ is define as $V^\pi(s) = \mathbb{E}_\pi[\sum_{k=0}^{\infty} \gamma^k R(s^{(k)}, \pi(s^{(k)}))|s_0 = s)]$. The team's goal is to compute the optimal value function V^* which is the fixed point of the Bellman equation $V^\pi(s) = \max_{a \in \mathcal{A}} Q(s, a)$ where the state-action value function Q is define as $Q(s, a) = R(s, a) + \gamma \sum_{s' \in \mathcal{S}} P(s'|s, a)V^\pi(s')$. The optimal policy, π^* is therefore compute using the definition $\pi^* \in \mathrm{argmax}_{a \in \mathcal{A}} Q(s, a)$. To compute such a solution, we employ the standard value iteration algorithm [19], which is proven to converge to a deterministic optimal policy in fully observable, finite-state, discounted settings [17].

4 Experiment Validation

4.1 Setup

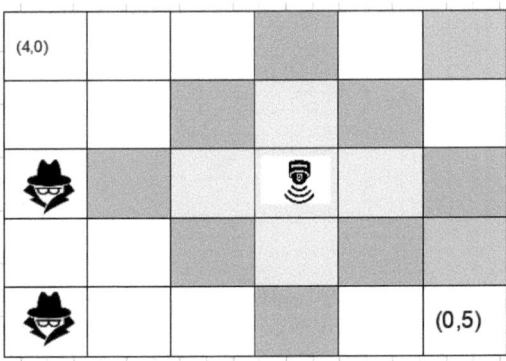

Fig. 1. 5×6 grid-world: Blue (p_1=0.8), Pink (p_2=0.25), Yellow (goals). (Color figure online)

We employed a 5×6 grid-world, depicted in Fig. 1, to enhance the interpretability of our results by highlighting the spatial dynamics of agent movements and their impact on network security. This setup helps illustrate how attackers can exploit vulnerabilities, similar to real-world threats. A state is defined by (row, col), with two attackers starting from $(0,0)$ and $(2,0)$. Each attacker can move in one of four directions: "N" (North or arrow up), "S" (South or arrow down), "E" (East or arrow right), or "W" (West or arrow left). When moving "N" the attacker has a probability of $1-2\alpha$ to reach the intended cell, while each adjacent cell (West and East) is entered with a probability of α. In our experiment, we set $\alpha = 0.1$. While its specific value is arbitrary, α enables calibration of the attack graph's probabilistic behavior to study the impact of uncertainty on security measures and attack outcomes. The environment features two yellow-colored goal states. The sensor is positioned at cell $(2,3)$, with the blue zone indicating high detection probability p_1, and the pink zone indicating a lower probability p_2 ($p_1 \geq p_2$). Outside these zones, agents are undetectable by the sensor. In addition, if multiple agents are close to a sensor, the mechanism detects the nearest one, or randomly selects among those equidistant. When several sensors are triggered, the system prioritizes the strongest detection signal; if multiple signals have equal intensity, one is selected at random.

Although our experiments are conducted in a grid-world, this environment serves as a generalizable benchmark and does not constrain the applicability of our framework to more complex networks [18],?.

4.2 Results and Discussion

To illustrate how policies adapt under different environmental settings, we first present the optimal policy for a single agent navigating the environment, both with and without detection mechanisms. This baseline policy serves to evaluate changes in agent's behavior based on the environment dynamics. Figure 2 highlight the impact of detection mechanisms on individual decision-making.

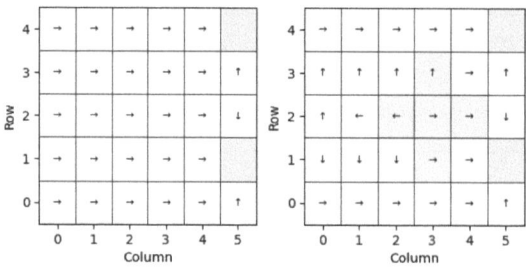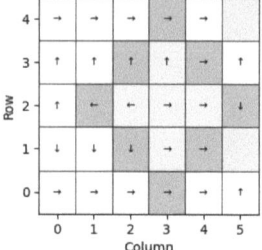

(a) agent's optimal move when there is no detection mechanism

(b) agent's optimal move when there is a detection mechanism with $p_1 = 0.8$ and $p_2 = 0$.

(c) agent's optimal move when there is a detection mechanism with $p_1 = 0.8$ and $p_2 = 0.25$.

Fig. 2. Single agent's optimal policy

We observe that in the absence of a detection mechanism, the agent mainly follows the right arrow's action to reach the goal states. However, when the detection mechanism is introduced, the agent attempts to reach the targets while avoiding entering the detection zones.

We evaluate three strategies for a gid with $p_1 = 0.8$ and $p_2 = 0.25$; coordinated (joint optimization), isolated (independent optimization), and myopic (greedy movement) for goal achievement. We launch 10000 episodes for each strategy and use for the coordinated policy, reward ratios $r = 1/10$ (balanced objectives) and $r = 1/1000$ (full-collection emphasis) to quantify coordination strictness. Although coordinated and isolated policies outperform myopic approaches (Fig. 3), their efficacy diverges based on objectives: For single-goal missions, isolated policies excel (85% vs. 65% success with $r = 1/10$, 64% with r=1/1000)) due to faster, uncoordinated execution that exploits early opportunities before sensor escalation. Conversely, coordinated strategies dominate multi-goal scenarios (30% with $r = 1/10$, 40% with $r = 1/1000$ vs 4%) by systematically staggering agent exposures to avoid simultaneous detections. This trade-off arises from the sensor escalation effect: isolated agents capitalize on rapid initial progress, whereas coordination prioritizes team survival at the cost of delayed individual success.

Coordinated strategies require more steps (12.42) to reach at least one target, compared to isolated (8.76) and myopic (7.98) strategies, and similarly for full target coverage (10.18 vs. 8.91 and 8.18). This reflects a deliberate trade-off: coordination prioritizes team safety over speed, making it better suited for missions where reaching all goals is essential, not just any.

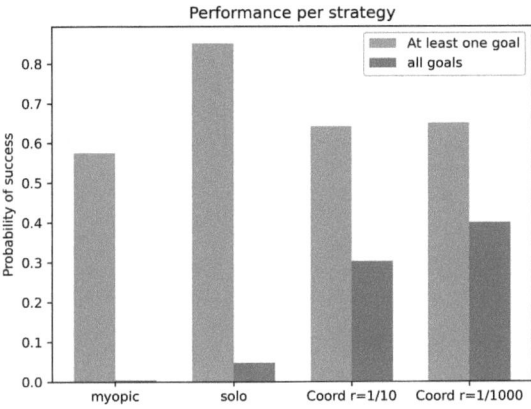

Fig. 3. Policy performance across mission objectives comparing: (1) myopic (greedy strategies), (2) isolated optimization (solo), and (3) coordinated strategies with reward ratios r = 1/10 (balanced single/full collection) vs r = 1/1000 (all-goal emphasis).

5 Conclusion

This paper presents a formal framework for modeling coordinated cyberattacks in networks with probabilistic detection mechanisms. By extending individual agent MDPs to an augmented Markov game, we capture the dynamic composition of a team of attackers subject to detection-induced neutralization. Our model incorporates realistic elements such as spatially dependent detection, implicit coordination through shared objectives, and reward structures reflecting trade-offs between partial and complete mission success.

Through grid-world simulations, we show that with our objective, uncoordinated (solo) policies can outperform coordinated strategies in simple reachability scenarios where success of a single agent suffices, while coordinated policies excel in missions requiring comprehensive target coverage. Overall, our contribution provides a principled framework to quantitatively analyze coordinated attackers under risk of detection, illustrating how progressive team attrition shapes optimal strategies. Even when some agents are neutralized, remaining attackers can still accomplish the mission, highlighting the need for defenders to anticipate partial success scenarios and design sensor placement, risk assessment, and defensive strategies accordingly.

We acknowledge that a primary limitation of the current approach is the exponential growth of the state and action space with the number of attackers, which restricts exact computation to small or medium-sized systems. Future work will explore scaling via approximate methods and learning-based strategies, integration of incomplete or deceptive sensor knowledge, adaptive detection mechanisms, and robust real-time defenses against dynamic multi-agent threats.

Acknowledgments. The research was sponsored by the U.S. Army Research Office and was accomplished under Cooperative Agreement Numbers W911NF-19-2-0150,

W911NF-22-2-0175, and Grant Number W911NF-21-1-0326. The views and conclusions contained in this document are those of the authors and should not be interpreted as representing the official policies, either expressed or implied, of the U.S. Army Research Laboratory or the U.S. Government. The U.S. Government is authorized to reproduce and distribute reprints for Government purposes notwithstanding any copyright notation herein.

References

1. Arikkat, D.R., et al.: SeCTIS: a framework to secure CTI sharing. Fut. Gener. Comput. Syst. **164**, 107562 (2025)
2. Bertoli, P., et al.: Representation and reasoning about coordinated attacks. In: AAMAS Workshops (2003)
3. Bitirgen, K., Filik, U.B.: Markov game based on reinforcement learning solution against cyber–physical attacks in smart grid. Expert Syst. Appl. **255**(PB) (2024)
4. Bleumer, G.: Threshold Signature. Springer, Living Reference Work (2023)
5. Boutilier, C.: Sequential optimality and coordination in multiagent systems. In: IJCAI, vol. 99, pp. 478–485 (1999)
6. Center, C.C.: Module 4-types of intruder attacks. CERT/CC Overview Incident and Vulnerability Trends (2003)
7. Chowdhary, A., et al.: Markov games for moving target defense. In: GameSec (2018)
8. Gao, P., Liu, P., et al.: Dynamic attack graphs with alert-driven reconfiguration. IEEE Trans. Depend. Sec. Comput. **123**, 102938 (2021)
9. Gennaro, R., Goldfeder, S., Narayanan, A.: Threshold-optimal DSA/ECDSA signatures and an application to bitcoin wallet security. Cryptology ePrint Archive, Report 2016/013 (2016)
10. Horák, K., Bošanský, B.: Game-theoretic approach to probabilistic attack graphs. Comput. Secur. **19** (2019)
11. Huang, Y., Xiong, Z., Zhu, Q.: Cross-layer coordinated attacks on cyber-physical systems: a LQG game framework with controlled observations. In: 2021 European Control Conference (ECC), pp. 521–528. IEEE (2021)
12. Kouam, W., Hayel, Y., Deugoue, G., Kamhoua, C.: Exploring centrality dynamics for epidemic control in complex networks: an asymmetrical centralities game approach. Dyn. Games Appl. **15**(3), 1–33 (2024)
13. Kouam, W., Hayel, Y., Kamhoua, C., Deugoué, G., Tsemogne, O.: A network centrality game based on a compact representation of defender's belief for epidemic control. J. Dynam. Games (2024)
14. Li, L., Ma, H., Han, S., Fu, J.: Synthesis of proactive sensor placement in probabilistic attack graphs. In: 2023 American Control Conference (ACC), pp. 3415–3421. IEEE (2023)
15. Na, G., Lee, H., Eun, Y.: A multiplicative coordinated stealthy attack for nonlinear cyber-physical systems with homogeneous property. Math. Probl. Eng. **2019**(1), 7280474 (2019)
16. Nyberg, J., Johnson, P.: Learning automated defense strategies using graph-based cyber attack simulations. In: Proceedings of WOSOC 2023: Workshop on Security Operation Center Operations and Construction, p. 4. Internet Society, San Diego, CA, USA (2023)

17. Puterman, M.L.: Markov Decision Processes: Discrete Stochastic Dynamic Programming. John Wiley & Sons (2014)
18. Sen, Ö., Pohl, C., Hacker, I., Stroot, M., Ulbig, A.: Ai-based attacker models for enhancing multi-stage cyberattack simulations in smart grids using co-simulation environments. In: 2024 IEEE International Conference on Cyber Security and Resilience (CSR), pp. 68–75. IEEE (2024)
19. Sutton, R.S., Barto, A.G.: Reinforcement Learning: An Introduction. MIT press (2018)
20. Wang, Z., Li, C.: Adversarial multi-agent coordination under partial observability. In: USENIX Security (2022)

Best-Response Learning in Budgeted α-Fair Kelly Mechanisms

Cleque Marlain Mboulou-Moutoubi[1]($\boxtimes$), Younes Ben Mazziane[1], Francesco De Pellegrini[1], and Eitan Altman[1,2]

[1] LIA, Avignon University, Avignon, France
`cleque-marlain.mboulou-moutoubi@univ-avignon.fr`
[2] INRIA, Sophia Antipolis, France

Abstract. The Kelly mechanism is a proportional allocation auction widely adopted in decentralized resource allocation systems to share an infinitely divisible resource among competing agents. We analyze the sequential game it induces when agents have α-fair utilities and behave strategically. Our main result proves that synchronous best-response updates drive bids to the unique Nash equilibrium at a linear rate for $\alpha \in \{0, 1, 2\}$. Extensive simulations reveal that best-response dynamics reach equilibrium significantly faster than previously proposed no-regret learning algorithms.

Keywords: decentralized resource allocation · auctions · game theory · Kelly mechanism · α-fair allocation

1 Introduction

Decentralized resource allocation problems arise in edge computing, cloud platforms, and communication networks [1–3]. Kelly [4] proposed a proportional allocation scheme in which each agent receives a share of the resource proportional to their bid. In that seminal work, the author made the explicit example of proportional fairness type of utilities. Later on, many extensions of the Kelly mechanism adopted the α-fair utility framework [5–7]. This class of utility functions is widely studied as they capture a range of trade-offs between fairness and efficiency: from efficiency-maximizing behavior ($\alpha = 0$), to proportional fairness ($\alpha = 1$), and approaching max-min fairness as $\alpha \to \infty$. α-fair utilities of the type studied in this work have since become a standard in modeling user preferences in network economics and decentralized optimization [7,8].

Johari and Tsitsiklis [9] proved that the game induced by the Kelly allocation mechanism admits a unique Nash equilibrium when utilities are smooth and concave, and that the social welfare at this equilibrium is at least 3/4 of the optimal. These guarantees, however, presuppose that agents can play the NE, which requires complete knowledge of every player's utility by all players. In this work we part from this setting and we assume instead that each agent knows

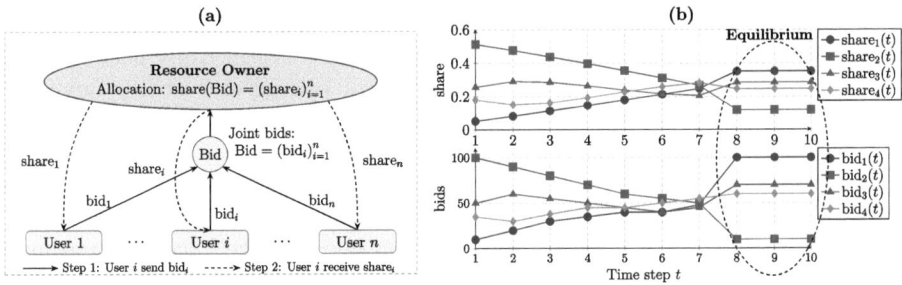

Fig. 1. From the static to the Sequential Kelly mechanism.

only their own utility. But, they can adapt their bid over repeated, synchronous rounds using limited feedback, e.g., the aggregate bid across all players.

Generally, in repeated games, the behavior of selfish agents is captured according to two main models, namely *best-response dynamics* (BRD) or *no-regret* learning algorithms. In best-response dynamics (BRD), each player exploit priors on opponents' actions and determines the best response action to be played at the next step accordingly [10]. No-regret learning algorithms are state of the art solutions for online convex optimization problems [11].

Mertikopoulos et al. [12] showed that Dual Averaging (DA) updates—a class of no-regret learning algorithms—converge to an NE whenever the underlying static game is *variationally stable*. Unfortunately, verifying variational stability is hard; it is known only for special cases of the Kelly game with linear utilities [13] or logarithmic utilities [14]. A further drawback of DA-based bidding schemes for the Kelly game is the lack of provable convergence *rates*. If DA converges slowly and the horizon is limited, agents may not reach the NE, undermining the social welfare guarantees that hold only at equilibrium.

Contributions. We show that if agents follow a simple synchronous best-response rule that uses only the previous round's aggregate bid, the system converges to the unique NE in *linear* time, provided that minimum-bid constraints are satisfied and agents utilities are of the α-fair form in the allocated resource for $\alpha \in \{0, 1, 2\}$. To our knowledge, this is the first linear-time convergence guarantee for the repeated Kelly game. Extensive simulations confirm that best-response dynamics reach equilibrium substantially faster than prior DA-based algorithms [13,14], preserving the welfare guarantees even in short horizons.

2 Problem Formulation

We consider a decentralized setting in which a unit-sized, divisible resource must be allocated among n strategic agents. These agents interact repeatedly in T rounds, each submitting a bid to request a share of the resource (see Fig. 1).

At each round, the resource owner assigns a share of the resource according to the Kelly mechanism. This mechanism extends the classical Kelly network optimization framework [4] by incorporating general pricing and utility functions, as in [15]. It allows for general α-fair utilities and non-linear bid-to-payment mappings.

In the following, we define the corresponding game $\mathcal{G}_\alpha$, namely the sequential Kelly mechanism with α-fair utility under budget constraints. We detail the players' action set, i.e., their bids, the allocation rule, i.e., the fraction of resources the resource owner assigns to each player given the bid vector, and the players' resulting utility. Each agent aims to maximize their own utility by signaling their request to the resource owner, subject to a budget constraint.

Bidding and Allocation. In the basic formulation of the Kelly mechanism [9], each user i submits a bid z_i to the resource owner; we assume $z_i > 0$. Let's denote by $\mathbf{z} = (z_1, \ldots, z_n)$ the profile of bids across all agents.

Given a strategy profile $\mathbf{z}$ and a system reservation parameter $\delta \geq 0$, the mechanism allocates to each agent i a fraction of the unit resource as follows:

$$x_i(\mathbf{z}) = \begin{cases} \frac{z_i}{\sum_{j=1}^n z_j + \delta}, & \text{if } z_i > 0, \\ 0, & \text{otherwise.} \end{cases} \quad (1)$$

The corresponding payment function is $p_i : \mathbb{R}_+ \mapsto \mathbb{R}_+$ defined as $p_i(z_i)$, and the budget constraint imposes $p_i(z_i) \leq c_i$, where $c_i > 0$ is agent i's private budget. Finally, each agent chooses their bid from the constrained strategy space:

$$\mathcal{R}_i = \{z_i \geq 0 \mid \epsilon_i \leq p_i(z_i) \leq c_i\}, \quad (2)$$

The strategy space of the game $\mathcal{G}_\alpha$ is thus $\mathcal{R} = \mathcal{R}_1 \times \cdots \times \mathcal{R}_n$.

Utility Function. From their assigned resource fraction $x_i = x_i(\mathbf{z})$, agent i obtains α-fair valuation function $V_i(\cdot)$

$$V_i(\mathbf{z}) = \begin{cases} \frac{x_i(\mathbf{z})^{1-\alpha}}{1-\alpha}, & \text{if } \alpha \neq 1, \\ \log(x_i(\mathbf{z})), & \text{if } \alpha = 1, \end{cases} \quad (3)$$

Also, each agent is assigned a valuation weight or scaling factor $a_i > 0$ which reflects the worth assigned by agent i for each unit of received resource. In the rest of the paper, we consider a linear cost model $p_i(z_i) = \lambda z_i$, where λ denotes the resource price determined by the resource owner. Consequently, their total payoff writes

$$\varphi_i(\mathbf{z}) = a_i V_i(\mathbf{z}) - \lambda z_i, \quad \forall i \in [n] \quad (4)$$

Let $\mathbf{z}_{-i} = (z_1, \cdots, z_{i-1}, z_{i+1}, \cdots, z_n)$ be the action of all players except i.

Definition 1 (Nash Equilibrium). *A strategy profile* $\mathbf{z}^* = (z_1^*, z_2^*, \ldots, z_n^*) \in \mathcal{R}$ *is a* Nash Equilibrium (NE) *if, for every player* $i \in [n]$,

$$\varphi_i(z_i^*, \mathbf{z}_{-i}^*) \geq \varphi_i(z_i, \mathbf{z}_{-i}^*) \quad \forall z_i \in \mathcal{R}_i \quad (5)$$

We observe that the function φ_i is concave in its i-th component and twice continuously differentiable on $\mathbb{R}_{>0}^n$, and the actions set $\mathcal{R}$ is non empty, closed, bounded, and convex. Thus, existence of a Nash Equilibrium (NE) of the game $\mathcal{G}_\alpha$ follows by [16, Theorem 1]. The minimum bid can be seen as a way to impose a reservation $\delta = \sum \epsilon_i$ and enforce minimal participation from all agents.

The next section presents the Best Response Dynamics (BRD), a simple action update rule that leads to equilibrium, with theoretical convergence guarantees in $\mathcal{G}_\alpha$.

3 Main Results

We study a system where each agent updates their strategy according to a Best-Response Dynamic (BRD) based on the latest actions of the other players. Formally, in a game with n players and strategy profile $\mathbf{z} \in \mathcal{R}$, the best response $\mathrm{BR}_i : \mathbb{R}_+^{n-1} \mapsto \mathbb{R}_+$ of agent i is defined as:

$$\mathrm{BR}_i(\mathbf{z}_{-i}) \in \arg\max_{z_i \in \mathcal{R}_i} \phi_i(z_i, \mathbf{z}_{-i}) \qquad (6)$$

In $\mathcal{G}_\alpha$, the utility function (4) of each agent depends only on their own bid and the aggregate bid of the others, making it an *aggregative game*. Consequently, the best-response update simplifies and depends only on the aggregate bid of the other agents, defined as $s_{-i} = \sum_{j \neq i} z_j + \delta$, so that $\mathrm{BR}_i(\mathbf{z}_{-i}) \triangleq \mathrm{BR}_i(s_{-i})$. Since $\mathcal{G}_\alpha$ is also a concave game, the best response is unique [17]:

$$\mathrm{BR}_i(s_{-i}) = \arg\max_{z_i \in \mathcal{R}_i} \varphi_i(z_i, \mathbf{z}_{-i}). \qquad (7)$$

We study the Synchronous Best Response Dynamic (SBRD), in which all agents simultaneously update their actions based on their best responses. The pseudocode is reported in Algorithm 1.

Algorithm 1. Synchronous Best Response Dynamic (SBRD)

1: **Input:** Initial bids $\{z_i(0)\}_{i=1}^n$, number of iterations T.
2: **Output:** Bid trajectory $\{\mathbf{z}(t)\}_{t=1}^T$.
3: **for** $t = 1$ to T **do**
4:     **for each** bidder $i \in [n]$ **simultaneously do**
5:         $s_{-i}(t-1) \leftarrow \sum_{j \neq i} z_j(t-1) + \delta$
6:         $z_i(t) \leftarrow \mathrm{BR}_i(s_{-i}(t-1))$ ▷ See Eq. 9
7:     **end for**
8: **end for**

Now, we analyze the convergence of SBRD(1) in the setting of $\mathcal{G}_\alpha$. The best response $\mathrm{BR}_i(s_{-i})$ of agent i, given the aggregate bid s_{-i}, is obtained by solving the first-order optimality condition $\frac{\partial \varphi_i(z_i, \mathbf{z}_{-i})}{\partial z_i} = 0$, and projecting the solution $\widetilde{BR}_i(s_{-i})$ onto the feasible set $\mathcal{R}_i$: $BR_i(s_{-i}) = \Pi_{\mathcal{R}_i}\left(\widetilde{BR}_i(s_{-i})\right)$, where $\Pi_{\mathcal{R}_i}(x) = \min\left\{\max\left\{x, \frac{\epsilon_i}{\lambda}\right\}, \frac{c_i}{\lambda}\right\}$. For $\alpha = 0$ the calculation is straightforward. For $\alpha > 0$, this condition yields a nonlinear equation that depends of the fairness parameter α and can be reformulated as:

$$\phi_{s_{-i}}(z_i, \alpha) = \underbrace{\left(\frac{z_i}{z_i + s_{-i}}\right)^\alpha}_{\phi^1_{s_{-i}}(z_i, \alpha)} - \underbrace{\frac{a_i s_{-i}}{\lambda(z_i + s_{-i})^2}}_{\phi^2_{s_{-i}}(z_i)} = 0 \tag{8}$$

where the solution $\widetilde{BR}_i(s_{-i})$ is given by the intersection of the functions $\phi^1_{s_{-i}}(\cdot, \alpha)$ and $\phi^2_{s_{-i}}(\cdot)$. However, solving this nonlinear equation for general $\alpha > 0$ is analytically intractable. Therefore, we derive explicit solutions for three key and representative cases $\alpha = 0, 1$, and 2 corresponding respectively to *efficiency maximization*, *proportional fairness*, and *minimum potential delay fairness* [18], and analyze how the solutions depend on agents' valuations and budgets. We identify further conditions to ensure convergence of SBRDs to a unique NE.

Lemma 1 (Best Response Operator). *Consider the game $\mathcal{G}_\alpha$ in the strategy space $\mathcal{R}$. The best response of player i to the strategy profile of the others, denoted $BR_i(s_{-i})$, is defined as*

$$BR_i(s_{-i}) = \Pi_{\mathcal{R}_i}(\widetilde{BR}_i(s_{-i})) \tag{9}$$

where $\widetilde{BR}_i(s_{-i})$ depends on the fairness parameter α. In particular, it holds

i. $\widetilde{BR}_i(s_{-i}) = \sqrt{\frac{a_i s_{-i}}{\lambda}} - s_{-i}$ *if* $\alpha = 0$; iii. $\widetilde{BR}_i(s_{-i}) = \sqrt{\frac{a_i s_{-i}}{\lambda}}$ *if* $\alpha = 2$

ii. $\widetilde{BR}_i(s_{-i}) = \frac{-s_{-i} + \sqrt{s_{-i}^2 + \frac{4 a_i s_{-i}}{\lambda}}}{2}$ *if* $\alpha = 1$

The explicit form of $\widetilde{BR}_i(s_{-i})$ is obtained by solving Eq. (8) given α. Here we state the convergence properties of the SBRD in $\mathcal{G}_\alpha$, under the representative values of α:

Theorem 1 (SBRD Convergence). *In the game $\mathcal{G}_\alpha$, the SBRD converges to the unique Nash equilibrium $\mathbf{z}^*$ of the Kelly mechanism if the minimum bid ϵ satisfies:*

1. *For $\alpha = 0$ and $n = 2$ if :* $\epsilon > \frac{\max(a_1, a_2)}{16\lambda} - \delta$
2. *For $\alpha = 1$ if :* $\epsilon > \frac{1}{\lambda\sqrt{n}(n-1)}(n - 2\sqrt{n} + 1) a_i^{\max} - \frac{\delta}{n-1}$
3. *for $\alpha = 2$ if :* $\epsilon > \frac{1}{4\lambda(n-1)}\left(\max_j \sum_{i \neq j} \sqrt{a_i}\right)^2 - \frac{\delta}{n-1}$

Moreover, the SBRD converges linearly to the Nash equilibrium of game $\mathcal{G}_\alpha$.

The idea for the proof, reported in Appendix 6 (see Proof of Theorem 1), is that the best-response operator 9 is in fact a contraction. More precisely, we prove that it is a contraction under the $\|\cdot\|_\infty$ norm for $\alpha = 0$ and $\alpha = 1$, whereas it is a contraction under $\|\cdot\|_1$ norm for $\alpha = 0$ and $\alpha = 2$. This is sufficient to grant convergence based on Banach's Fixed Point Theorem. Incidentally, this also grants the uniqueness of the Nash equilibrium, i.e., of the fixed point. The contraction argument used in the proof also brings the speed of convergence of the best-response dynamics, namely: **Remarks:** (a) Theorem 1 requires $\epsilon = o(1/n)$ for $\alpha = 1$. In practice, as shown in the next section, ϵ can be set to a small value for a small number of players without hindering the convergence of the SBRD. (b) Classical results on the convergence of the BRD rely on the existence of a potential [19]. However, for the Kelly mechanism, it can be shown that a potential does exist if and only if $\alpha = 1$ (we omit here the proof for the space's sake). Nevertheless, in our framework, it is immediate to see that the *unilateral* best-response dynamics also converge, under the same assumptions.

4 Numerical Simulations

In this section we compare convergence properties of the SBRD with two established decentralized learning algorithms adapted to the Kelly mechanism. The comparison is performed for the three reference fairness regimes, namely $\alpha = 0, 1, 2$. The two algorithms are the following ones.
(i) **Dual Averaging with Quadratic Regularizer (DAQ)** [14,20], each player i maintains a *cumulative discounted gradient* of their utility $g_i(t)$, define as: $g_i(t) = \sum_{k=0}^{t-1} \eta_k \partial_i \varphi_i(z_i(k))$, with η_k being the step size at time k. The bid $z_i(t)$ is updated at time t by projecting $g_i(t)$ using $\Pi_{\mathcal{R}_i}$;
(ii) **Exponential Learning (XL)** [13], where each player i uses $g_i(t)$ to update their bid via a sigmoid mapping projected on action space: $z_i(t) = \Pi_{\mathcal{R}_i}(c_i \sigma(g_i(t)))$ with $\sigma(x) = \frac{1}{1+\exp(-x)}$ and c_i is the budget.

Convergence to the unique Nash equilibrium in both DAQ and XL is ensured under DSC condition [16, Theorem 2], already proved for the case $\alpha = 1$ for DAQ (see [14, Lma. 2]) and for $\alpha = 0$ for XL [13, Thm. 2]. The request for the step size η_t is usually $\sum_{t=1}^\infty \eta_t = +\infty$ and $\sum_{t=1}^\infty \eta_t^2 < +\infty$. However, we use a fixed step-size η_t to speed up the convergence of those algorithms [13,14,20].

Convergence is measured by tracking the best response residual $\|\text{BR}(\mathbf{z}(t)) - \mathbf{z}(t)\|$, that is the distance of the current multi-strategy vector from the NE. In all tests, convergence is attained for a best response residual of 10^{-5}.

The experiments compare the convergence trajectories of the three methods, focusing on how the fairness parameter α and the update strategy affect convergence speed and stability in a game with heterogeneous players.

The game consists of $n \geq 2$ agents, $i = 1, 2, \ldots, n$ with utility function as in (4), where $a_i = 100 \cdot i^{-\gamma}$ and a fixed price $\lambda = 1$. We note that increasing γ introduces larger heterogeneity among players; $\gamma = 0$ represents homogeneity. Each agent has a limited budget $c_i = 4 \times 10^3$. The minimum bid is set to $\epsilon_i = 10^{-3}$ and $z_i(0) = 1$ is the initial bid. The reservation parameter in the allocation function is $\delta = 0.1$.

Proportional Fairness ($\alpha = 1$): Figs. 2a, 2b, 2c. In the proportional fairness regime, SBRD consistently achieves the fastest convergence, driving the system to the NE with less than **10 iterations**. XL is slower and convergence occurs in the order of 500 steps. We observe that its convergence speed is not affected by the players' utility heterogeneity, i.e., different values of γ. In contrast, DAQ requires more than 1500 iterations to to converge.

Minimum Potential Delay Fairness ($\alpha = 2$): Figs. 2d, 2e, 2f. As in the previous case, in this fairness regime SBRD shows a similar convergence speed, for all tested values of γ. Here, it converges in **less than 30 iterations**. In contrast, XL converges in more than 2000 steps and DAQ beyond 10^5 steps (even for decreasing stepsize). We observe that, as for $\alpha = 1$, both XL and DAQ appear to converge faster for larger values of γ, i.e., for more pronounced heterogeneity in the players' payoff.

Efficiency Maximization ($\alpha = 0$): Fig. 2g. In the efficiency-maximizing regime, the convergence of SBRD, DAQ, and XL is guaranteed for $n = 2$. Again, SBRD reaches the NE significantly faster than DAQ and XL. When $n > 2$, the convergence of SBRD is not theoretically established, but we can further evaluate it empirically. Figure 2g shows that SBRD maintains consistent performance when agents are heterogeneous (i.e., $\gamma > 0$), for the tested size of the game $n = 2, 3, 4, 20$. However, as we could somehow expect, SBRD fails to converge in highly homogeneous environments with $n \geq 4$, confirming some limitations in its applicability in this regime in a homogeneous setting.

General Case ($\alpha > 0$): Fig. 2h. We want to test the performance of SBRD for general values of $\alpha > 0$, where the BR can be evaluated numerically using a simple bisection search. Figure 2hshows the SBRD time to convergence, corresponding to a value 10^{-5} for the BR residual. Here, $c_i = 3 \cdot 10^4$ and $\gamma = 0.4$. For $\alpha \geq 1/2$, the SBRD converges rapidly across all game sizes. It appears to have a minimum with convergence time decreasing from $\alpha = 0.5$ to 1, and increasing in a linear fashion beyond.

5 Conclusion

We have explored the decentralized proportional resource allocation, commonly known as the Kelly mechanism, within the context of α-fair utility functions, and subject to budget limitations. Under mild assumptions on the minimum bid, we demonstrated linear convergence of the synchronous best-response dynamics for several cases of interest. We evaluated its performance against dual averaging algorithms. Our findings indicate that, within its stability region, SBRD converges at a markedly faster rate when compared with no-regret algorithms that often require a large number of iterations to achieve equilibrium. In particular, in our numerical results we could confirm the convergence of SBRD for $\alpha > 1/2$. In that region, in sight of its speed of convergence, SBRD appears the natural option to obtain a lightweight sequential scheme based on the Kelly mechanisms under α-fair utilities.

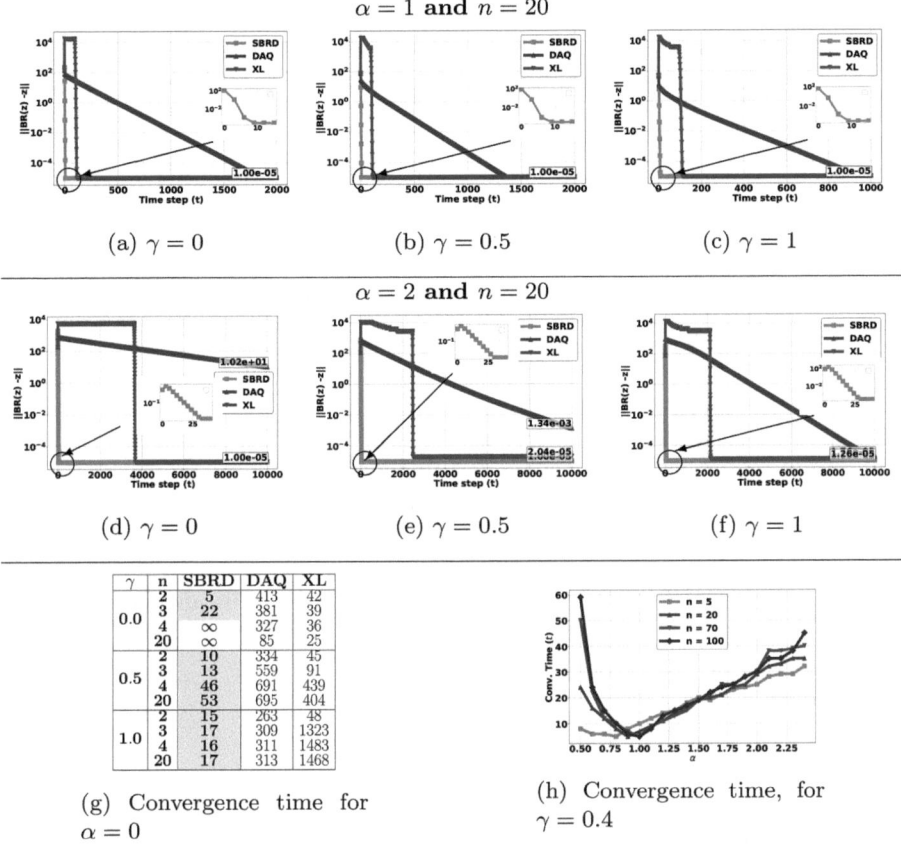

Fig. 2. SBRD Convergence as a function of γ for different values of α.

In future works, we plan to extend our analysis in order to define analytically the region of convergence of SBRD with respect to fairness parameter α. Another interesting direction entails the presence of multiple coupled resources for which players need to bid. This would require to study a multi-dimensional version of the Kelly mechanism and the consequent extension of the SBRD to this case.

6 Appendix: Proof of Theorem 1

Proof. In order for the best response operator to be a contraction on $\mathcal{R}$ we need to identify $0 < q < 1$ such that

$$\|\text{BR}(\boldsymbol{z_1}) - \text{BR}(\boldsymbol{z_2})\| \leq q \|\boldsymbol{z_1} - \boldsymbol{z_2}\|$$

for all $\boldsymbol{z_1}, \boldsymbol{z_2} \in \mathcal{R}$.

From the generalized mean value theorem (see [21][Cor. 3.2]), it holds

$$\|\text{BR}(\boldsymbol{z_1}) - \text{BR}(\boldsymbol{z_2})\| \leq \sup_{\boldsymbol{z} \in \mathcal{R}} \|\mathcal{J}_{\text{BR}}(\boldsymbol{z})\| \|\boldsymbol{z_1} - \boldsymbol{z_2}\|$$

so that the BR operator is a contraction if $q := \sup_{z \in \mathcal{R}} \|\mathcal{J}_{BR}(z)\| < 1$. Note that vector and operator norms need to be chosen consistently.

Note that BR converges linearly, with convergence rate determined by $q = \|\mathcal{J}_{BR}(S_0)\|$ [22, Thm. 4.2.1]. We thus focus on proving that the norm of the Jacobian matrix is strictly smaller than 1.

Since $\mathcal{G}_\alpha$ is an aggregative game 7, the Jacobian satisfies $(\mathcal{J}_{BR})_{i,j}(z) = (\mathcal{J}_{BR})_{i,j}(s_{-i})$. Let $S_0 = (s_0)_{i=1}^n$ where $s_0 = (n-1)\epsilon + \delta$ the smallest aggregated bid.

1. **Case $\alpha = 1$:** Use $\|\cdot\|_\infty$ norm. The Jacobian entries are: $(\mathcal{J}_{BR})_{i,j}(s_{-i}) = \begin{cases} \frac{-1}{2} + \frac{s_{-i} + \frac{2a_i}{\lambda}}{2\sqrt{s_{-i}^2 + \frac{4a_i s_{-i}}{\lambda}}} \geq 0, & j \neq i, \\ 0, & j = i. \end{cases}$, $(\mathcal{J}_{BR})_{i,j}(\cdot)$ is decreasing,

   Hence, $\|\mathcal{J}_{BR}(S_0)\|_\infty = \sup_{z \in \mathcal{R}} \|\mathcal{J}_{BR}(z)\|_\infty = \max_i \sum_j (\mathcal{J}_{BR})_{i,j}(s_0)$. Imposing $\|\mathcal{J}_{BR}(S_0)\|_\infty < 1$ leads to a quadratic inequality, $\lambda^2 s_0^2 + 4a_i^{\max} \lambda s_0 - \frac{(a_i^{\max})^2 (n-1)^2}{n} > 0$, satisfied for: $s_0 > (n - 2\sqrt{n} + 1) \frac{a_i^{\max}}{\lambda \sqrt{n}}$.

2. **Case $\alpha = 2$:** Use $\|\cdot\|_1$ norm. The Jacobian entries are:
   $(\mathcal{J}_{BR})_{i,j}(s_{-i}) = \begin{cases} \frac{1}{2}\sqrt{\frac{a_i}{\lambda s_{-i}}} \ (>0), & j \neq i, \\ 0, & j = i. \end{cases}$, $(\mathcal{J}_{BR})_{i,j}(\cdot)$ is also decreasing.

   The contraction condition becomes: $\|\mathcal{J}_{BR}(S_0)\|_1 = \max_j \sum_{i \neq j} \frac{1}{2}\sqrt{\frac{a_i}{\lambda s_0}} < 1$, implying: $s_0 > \frac{1}{4\lambda} \left(\max_j \sum_{i \neq j} \sqrt{a_i} \right)^2$.

3. **Case $\alpha = 0$:** Use $\|\cdot\|_\infty$ norm. The Jacobian satisfies:
   $(\mathcal{J}_{BR})_{i,j}(s_{-i}) = \begin{cases} -1 + \frac{1}{2}\sqrt{\frac{a_i}{\lambda s_{-i}}}, & j \neq i, \\ 0, & j = i, \end{cases}$

   leading to: $\|\mathcal{J}_{BR}(z)\|_\infty = (n-1) \max_i \left| -1 + \frac{1}{2}\sqrt{\frac{a_i}{\lambda s_{-i}}} \right| < 1$. Solving the inequality yields: $\frac{a_i}{s_{-i}} < 16\lambda$, $\forall i \in \{1, 2\}$. Setting a_i to be $\max(a_1, a_2)$ and s_{-i} to s_0 leads to constraints on ϵ that depend on δ and $\max(a_1, a_2)$.

This finishes the proof. □

References

1. Maillé, P., Tuffin, B.: Multi-bid auctions for bandwidth allocation in communication networks. In: *Proceedings IEEE INFOCOM 2004, The 23rd Annual Joint Conference of the IEEE Computer and Communications Societies, Hong Kong, China, March 7-11, 2004*. IEEE (2004)
2. Datar, M., Altman, E., Pellegrini, F.D., Azouzi, R.E., Touati, C.: A mechanism for price differentiation and slicing in wireless networks. In: *18th International Symposium on Modeling and Optimization in Mobile, Ad Hoc, and Wireless Networks, WiOPT 2020, Volos, Greece, June 15-19, 2020*, pp. 121–128. IEEE (2020)

3. Wang, X., Sui, Y., Wang, J., Yuen, C., Weiwei, W.: A distributed truthful auction mechanism for task allocation in mobile cloud computing. IEEE Trans. Serv. Comput. **14**(3), 628–638 (2021)
4. Kelly, F,P.: Charging and rate control for elastic traffic. *Europ. Trans. Telecommun.* **8**(1), 33–37 (1997)
5. Mo, J., Walrand, J.: Fair end-to-end window-based congestion control. IEEE/ACM Trans. Netw. **8**(5), 556–567 (2000)
6. Huaizhou, S.H.I., et al.: Fairness in wireless networks: issues, measures and challenges. *IEEE Commun. Surv. Tutor.* **16**(1), 5–24 (2013)
7. Srikant, R., Ying, L.: *Communication networks: an optimization, control and stochastic networks perspective.* Cambridge University Press (2014)
8. Kelly, F.P., Maulloo, A.K., Tan, D.K, H.: Rate control for communication networks: shadow prices, proportional fairness and stability. *J. Operat. Res. Soc.* **49**(3), 237–252 (1998)
9. Johari, R., Tsitsiklis, J.N.: Efficiency loss in a network resource allocation game. *Math. Oper. Res.* **29**(3), 407–435 (2004)
10. Lei, J., Shanbhag, U.V., Pang, J.-S., Sen, S.: On synchronous, asynchronous, and randomized best-response schemes for stochastic NASH games. Math. Oper. Res. **45**(1), 157–190 (2020)
11. Hazan, E.: Introduction to online convex optimization. *Found. Trends® Optim.* **2**(3-4), 157–325 (2016)
12. Mertikopoulos, P., Zhou, Z.: Learning in games with continuous action sets and unknown payoff functions. Math. Program. **173**(1–2), 465–507 (2019)
13. D'Oro, S., Galluccio, L., Mertikopoulos, P., Morabito, G., Palazzo, S.: Auction-based resource allocation in openflow multi-tenant networks. Comput. Netw. **115**, 29–41 (2017)
14. Mazziane, Y.B., Mboulou-Moutoubi, C.M., Pellegrini, F.D., Altman, E.: Learning to bid in proportional allocation auctions with budget constraints. In: *Proc. of IEEE WiOpt*, Linkoping, Sweden, May 26-29 (2025)
15. Maheswaran, R.T., Başar, T.: Efficient signal proportional allocation (ESPA) mechanisms: decentralized social welfare maximization for divisible resources. IEEE J. Sel. Areas Commun. **24**(5), 1000–1009 (2006)
16. Rosen, J. B.: Existence and uniqueness of equilibrium points for concave N-person games. *Econometrica* **33**(3) (1965)
17. Pellegrini, F.D., Massaro, A., Goratti, L., El-Azouzi, R.: Bounded generalized Kelly mechanism for multi-tenant caching in mobile edge clouds. In: *Network Games, Control, and Optimization: Proceedings of NETGCOOP 2016, Avignon, France*, pp. 89–99. Springer (2017)
18. Tang, A., Wang, J., Low, S.H.: Is fair allocation always inefficient. In: *IEEE INFOCOM 2004*, vol. 1. IEEE (2004)
19. Monderer, D., Shapley, L.S.: Potential games. *Games Econ. Behav.* **14**(1), 124–143 (1996)
20. Mertikopoulos, P., Zhou, Z.: Learning in games with continuous action sets and unknown payoff functions. *Math. Prog.* **173** (2019)
21. Coleman, R.: *Calculus on Normed Vector Spaces.* Springer (2012)
22. Kelley, C,T.: *Iterative Methods for Linear and Nonlinear Equations.* SIAM (1995)

The Effect of Network Topology on the Equilibria of Influence-Opinion Games

Yigit Ege Bayiz[✉][iD], Arash Amini[iD], Radu Marculescu[iD], and Ufuk Topcu[iD]

The University of Texas at Austin, Austin, TX 78712, USA
egebayiz@utexas.edu

Abstract. Online social networks exert a powerful influence on public opinion. Adversaries weaponize these networks to manipulate discourse, underscoring the need for more resilient social networksquery. To this end, we investigate the impact of network connectivity on Stackelberg equilibria in a two-player game to shape public opinion. We model opinion evolution as a repeated competitive influence-propagation process. Players iteratively inject *messages* that diffuse until reaching a steady state, modeling the dispersion of two competing messages. Opinions then update according to the discounted sum of exposure to the messages. This bi-level model captures viral-media correlation effects omitted by standard opinion-dynamics models. To solve the resulting high-dimensional game, we propose a scalable, iterative algorithm based on linear-quadratic regulators that approximates local feedback Stackelberg strategies for players with limited cognition. We analyze how the network topology shapes equilibrium outcomes through experiments on synthetic networks and real Facebook data. Our results identify structural characteristics that improve a network's resilience to adversarial influence, guiding the design of more resilient social networks.

Keywords: Network Resilience · Opinion Dynamics · Influence Propagation · Stackelberg Equilibrium

1 Introduction

Withstanding adversarial influence in social networks is a pressing issue across domains ranging from marketing to online influence campaigns [10,20]. In such settings, two competing players often engage in information conflict, each injecting messages into the network to influence public opinion. One player, the *adversary*, strategically spreads a harmful or misleading message, while the other, the *defender*, counters with a corrective or opposing message. This competition unfolds over time as each player sequentially inject messages into the network, aiming to influence the distribution of public opinion in their favor. Social networks evolve continually; as public opinion changes, individual connections adapt accordingly [9,12]. Consequently, to prevail in these information conflicts, players must not only maximize their immediate influence but also strategically shape the network—through targeted opinion distribution—to sustain their advantage over time.

Accurately modeling this competitive opinion evolution is therefore crucial for understanding network resilience against adversarial influence [1,7,8]. However, existing models of opinion dynamics and influence propagation largely fail to capture this multi-round competition, often considering only a single contagion spreading in isolation [4,13], or assuming static, non-adaptive influence sources [24]. In contrast, real-word influence conflicts involve co-evolving opinions and the game-theoretic interplay between the long-term decision-making of adversaries and defenders.

This paper introduces a *bi-level* competitive opinion update model to study conflicts in information environments, integrating a lower-level diffusion process for modeling competing influences and a top-level opinion update model. Formally, we consider a repeated sequential game between two players. At each round, the adversary (*leader*) introduces an opinionated *message* into the network, originating from a chosen point in the opinion space. The defender (*follower*) then responds by injecting a competing message. These *messages* then propagate through a competitive diffusion process. After diffusion, each individual in the network is predominantly influenced by either the defender or the adversary, reflecting the realistic constraint that individuals often choose between conflicting pieces of information. Once the influence diffusion process reaches an equilibrium state, the *opinions* of the individual update based on which player has the predominant influence over the individual. A key advantage of this bi-level approach is its ability to capture correlation effects message exposure and opinion evolution that are often missed by standard opinion dynamics models, such as the Friedkin-Johnsen type [6,11]. These standard models typically assume that an individual's opinion is simultaneously affected by all neighbors—subject to connection weights—irrespective of the specific influence each neighbor propagates at a given time. We assume that each player makes decisions according to the Stackelberg feedback equilibrium [18].

In summary, our key contributions are threefold. First, we introduce a novel bi-level influence-opinion model that formalizes the long-term opinion effects arising from the interplay between a malicious influence source and a defensive counter-influence source. Second, we formulate this attacker-defender interaction as a Stackelberg game and derive scalable, approximate local-feedback Stackelberg equilibrium solutions. Finally, we investigate how the homophily of social networks influences the resilience of public opinion to adversarial influence.

2 Problem Statement

Adversaries are increasingly exploiting social networks, creating a competitive dynamic in which defenders must counteract adversarial influence aimed at manipulating public opinion [5,19,23]. However, existing models often overlook the long-term nature of these campaigns and the co-evolution of beliefs and network, typically focusing on short-term influence spread or isolated dynamics

[2,13,24]. To bridge this gap, we model the adversarial competition as a bi-level influence-opinion Stackelberg game, where the adversary acts as the leader. This paradigm uniquely integrates rapid message propagation (influence spread) with the slower-paced evolution of public opinions and social ties, leveraging time-scale separation to better understand how persistent exposure shapes long-term beliefs and ultimately, network resilience against such manipulation.

Consider a network of n individuals that interact according to the time-varying communication graph $\mathcal{G}_t = \{V, W_t\}$. Here, $V := \{1, \ldots, n\}$ represents the set of individuals, and $W_t \in [0,1]^{n \times n}$ stores the probabilities of interaction between individuals at time t. Each individual $i \in V$ has an opinion $x_t^i \in \mathbb{R}^d$ in macro-time t. We assume that the underlying interactions are correlated with the individual's opinion. Specifically, the probability of individuals $i, j \in V$ interacting is correlated by $\psi(x_t^i, x_t^j)$, where $\psi(\cdot, \cdot) : \mathbb{R}^d \times \mathbb{R}^d \to [0,1]$ is the interaction kernel. The $[W_t]_{ij}$ is the row normalization of the matrix of interaction kernels, i.e.,

$$[W_t]_{ij} = \frac{\psi(x_t^i, x_t^j)}{\sum_{k \in V/i} \psi(x_t^i, x_t^j)}, \quad \forall\, i \neq j \in V, \tag{1}$$

and $[W_t]_{ii} = 0$. In each macro-time step t, the adversary initiates by sharing messages that support the opinion $u_t^a \in \mathbb{R}^d$. This is equivalent to strategically spreading messages from network nodes with opinions similar to u_t^a, using homophily-based connections. The defender then responds by choosing a message with opinion $u_t^d \in \mathbb{R}^d$ to counter the adversarial messages. We assume that the adversary aims to shift the opinions of individuals toward a target state x_a (known to both players), while the defender strives to maintain the status quo. Following initial message selections, individuals spread information rapidly across the network through the communication graph $\mathcal{G}_t$. Given that messages propagate significantly faster than opinions evolve, we model information propagation by linear dynamics, with another time scale, $s > 0$, denoted by micro-time.

The message of the defender aims to counter the adversarial influence. Therefore, we assign an *evidential value* of 1 to the defender's message and -1 to the adversary's message. In this way, we can abstract the propagation of opposing information. If an individual simultaneously receives adversarial and defender content, their evidential values cancel each other out. Assume that at some micro-time s within macro-time t, individual i has already accumulated evidential information $y_{t,s}^i \in \mathbb{R}$. Individuals share their current evidential information with a probability α, and the evidential information for the individual i evolves by

$$y_{t,s+1} = \alpha W_t y_s - p_t^a e^{-\kappa_a s} + p_t^d e^{-\kappa_d s} \tag{2}$$

Here, $[p_t^\square]_i = \psi(u_t^\square, x_t^i)$ represents the probability that i observes message shared by the adversary(or defender), based on their distance of opinions, $\square = a$ (or $\square = d$). We incorporate an exponential decay for the probability of interaction to model the community's diminishing interest in the messages over time. To maintain heterogeneity, we assume that the rate of interest decay differs for adversarial (κ_a) and defender (κ_d) messages.

We calculate the overall evidential information each individual receives during these interactions by

$$\bar{y}_t := \sum_{s=0}^{\infty} y_{t,s} = (I - \alpha W_t)^{-1} \left[p_t^d \frac{e^{-\kappa_d}}{1 - e^{-\kappa_d}} - p_t^a \frac{e^{-\kappa_a}}{1 - e^{-\kappa_a}} \right]. \quad (3)$$

Here, $\bar{y}$ represents the total accumulated evidence for each individual. Individuals then shift their opinions towards the defender's content u_t^d with a probability $\varsigma(\bar{y}^i)$, where $\varsigma(\cdot) : \mathbb{R} \to [0, 1]$ is a sigmoid function, and towards the attacker's content u_t^a with a probability $1 - \varsigma(\bar{y}^i)$. The individual opinions evolevs by

$$x_{t+1}^i = (1 - \lambda) x_0^i + \lambda \Big(x_t + \eta |\bar{y}_t^i| \big(\varsigma(\bar{y}_t^i) u_t^d + (1 - \varsigma(\bar{y}_t^i)) u_t^a - x_t \big) \Big), \quad (4)$$

where x_0^i denotes the initial opinion of individual i at time 0. This opinion evolution model is derived from the Friedkin-Johnson opinion dynamics model, with a learning rate $\eta |\bar{y}_t^i|$. In this model, individuals exhibit a degree of stubbornness toward their initial opinions, represented by λ. Note that learning rate reflect the amount of evidential information individual receives.

We assume that both the adversary and the defender aim to minimize a quadratic cost function. We heavily penalize sharing extreme ideological content, especially for the defender, who needs to remain neutral. For a given adversary policy $\pi_t^a(x)$, the defender seeks to solve the optimization problem

$$\max_{u_t^d \sim \pi_{1:T}^d} J_d := \sum_{t=1}^{T} \left[(x_t - x_0)^T Q_d (x_t - x_0) + u_t^{d^T} R_d u_t^d \right] \quad (5)$$

$$\text{s.t.} \quad x_{t+1} = F(x_t, \pi_t^a(x_t), \pi_t^d(x_t)),$$

where $F(\cdot, \cdot, \cdot)$, summarize the evolution of dynamics for the individuals. Solving (5) yields the optimal defender policy $\pi_t^{*d}(x_t, \pi_t^a)$. The adversary then solves

$$\max_{\pi_{1:T}^a} J_a := \sum_{t=1}^{T} \left[(x_t - \hat{x})^T Q_a (x_t - \hat{x}) + u_t^{a^T} R_a u_t^a \right] \quad (6)$$

$$\text{s.t.} \quad x_{t+1} = F(x_t, \pi_t^a(x_t), \pi_t^{*d}(x_t, \pi_t^a)).$$

Social networks often involve millions of people within a vast opinion space. Even in focused scenarios where the opinion space is reduced to a few dimensions, the number of individuals can still be a few thousand. This immense state space presents a significant challenge in solving the described optimization problem. In the following section, we introduce a dynamic clustering-based approach designed to efficiently address this challenge.

3 Methodology

In this section, we describe a scalable method to approximately compute a local feedback Stackelberg equilibrium. Our approach relies on dynamic clustering of the population to reduce the dimensionality of the problem. We then use a model-predictive-control approach to find an approximate local feedback Stackelberg equilibrium for the problem.

Dynamic Clustering. Real-world social networks often encompass millions of individuals, rendering the direct computation of local-feedback Stackelberg equilibria intractable due to their large scale. To address this curse of dimensionality, our model represents the individual population as a dynamic set of time-varying clusters, with assignments updated online. This approach enables operation on a *reduced graph* of connected clusters, the order of which is significantly smaller than the node count of the original social network.

In the initial macro-time step we construct the reduced graph by hierarchically clustering the network $\mathcal{G}$ into m_0, and building the quotient graph $\hat{\mathcal{G}}_0 = (\hat{\mathcal{S}}_0, \hat{W}_0)$, where $\hat{\mathcal{S}}_t = \{s_t^1, \ldots s_t^{m_t}\}$ is the set of clusters in macro-time t. We then treat each cluster as an individual and find W_t according to (1).

Rather than re-clustering at each time step, which can be computationally burdensome, we compute the cluster assignments $\hat{f}_t$ in all subsequent time steps by sequentially applying splitting and merging procedures on the clusters $\hat{f}_{t-1}$ in the previous time step. This process aims to split bimodal clusters into unimodal cluster pairs and merge pairs of clusters with unimodal mixture distributions into a single cluster.

Cluster Splitting. The splitting procedure iteratively splits clusters with sufficiently high bimodality into individual clusters by running a hierarchical clustering algorithm over the clusters with a fixed target cluster size of 2. We determine the splitting threshold for each cluster by tracking the skewness, γ_1, and kurtosis, γ_2, for the opinion distribution across the principal axis of each cluster and calculating the Sarle's bimodality coefficient, $\text{BC} = (\gamma_1^2 + 1)/\gamma_2$. We then compare this coefficient to a threshold value and split the cluster if the bimodality coefficient exceeds the threshold value. For our experiments, we found that a threshold of 0.55 works well and leads to stable cluster counts.

Cluster Merging. The merging rule we employ aims to ensure that all clusters S_t^i in the network have unimodal distributions. For each cluster i, we first compute the mean μ_i and the covariance matrix Σ_i. We then iterate over cluster pairs and greedily merge each pair if the means of each cluster in the pair lie within one standard deviation away from the other cluster's mean opinion. That is, we merge clusters i, j whenever the following condition holds for $d = \mu_i - \mu_j$,

$$||d||^2 < \min\left(\frac{d^\top \Sigma_i d}{d^\top d}, \frac{d^\top \Sigma_j d}{d^\top d}\right). \tag{7}$$

The above is a sufficient condition that guarantees the mixture distribution of the clusters i and j is unimodal, given that clusters i and j each have normal distributions [3]. Clearly, this assumption is not guaranteed to hold in the real-world networks; however, in practice, we observe that the competition between the players causes sufficient mixing for the cluster distributions to tend towards normal distributions.

Bounded Cognition Stackelberg Equilibrium. Recognizing that in practice decision-makers in this game are human experts whose behavior often deviates from the perfect rationality assumed in traditional game theory, our methodology departs from the perfect Stackelberg equilibrium concept [14,17]. Instead, we address the complexities of human cognition alongside the inherent high-dimensionality and nonlinearity of the problem by employing an iterative approach. In this approach, players model their opponent as possessing a cognitive capacity one level lower. To effectively manage persistent non-linearity, we first approximate the evolution of opinion cluster centers via linearization. Subsequently, we determine the feedback policy for each player by sequentially solving the resultant time-varying linear quadratic regulator (LQR) problem, based on the assumption that the opposing player operates according to this one-level-lower cognitive model [21]. We implement a receding-horizon control approach to model the decision making of the players. We assumes that players update their policies and cluster formations by observing the actual evolution of the social network after a period of policy application. This strategy is particularly well-suited for the inherent unpredictability and high uncertainty of social network dynamics, where reliable predictions are typically confined to short horizons.

Let $s_t \in \mathbb{R}^m$ denote the position of the centers of the clusters in some macro-time t, known to both players. Assume that we have calculated a reference trajectory, $\hat{s}^0_{t:t+H}$ as an initial estimate of the trajectory for the next H steps, calculated by assuming both the adversary and the defender play fixed policy $u^{a,0}_{t:t+h} = u^a_{t-1}$ and $u^{d,0}_{t:t+h} = u^d_{t-1}$ over the horizon. A first-order expansion of dynamics around the reference trajectory gives the perturbation dynamics

$$\bar{s}^0_{t+1} = A^0_t \bar{s}_t + B^{a,0}_t u^a_t + B^{d,0}_t u^d_t, \tag{10}$$

$$A^0_t = \left.\frac{\partial F}{\partial s}\right|_{(\hat{s}^0_t, \hat{u}^{a,0}_t, \hat{u}^{d,0}_t)}, \qquad B^\square_t = \left.\frac{\partial F}{\partial u^\square}\right|_{(\hat{s}^0_t, \hat{u}^{a,0}_t, \hat{u}^{d,0}_t)}, \qquad \square \in \{a,d\}.$$

Let us denote the feedback policy for the defender and attacker at the cognition level ℓ by $u^{d,\ell}_{t:t+H}(s)$, $u^{d,\ell}_{t:t+H}(s)$. Since the $t-1$ messages are available to both players, they both have an approximation of the other player zeroth cognition level input. At each cognition level the defender (*follower*) needs to solves

$$\min_{\{u^{d,l}_\tau\}^{t+H-1}_{\tau=t}} \sum_{\tau=t}^{t+H-1} \left[(\bar{s}^{\ell-1}_\tau)^T Q_d \bar{s}^{\ell-1}_\tau + (u^{d,\ell-1}_\tau)^T R_d u^{d,\ell-1}_\tau \right] \tag{8}$$

$$\text{s.t.} \qquad \bar{s}^{\ell-1}_{t+1} = (A^{\ell-1}_t + B^{a,\ell-1}_t K^{a,\ell-1}_t)\bar{s}^{l-1}_t + B^{d,\ell-1}_t u^{d,\ell-1}_t.$$

We can solve the optimization problem (8) by standard Riccati recursion to find the feedback for modified state matrix $\hat{A}^{d,l}_t = A^{\ell-1}_t + B^{a,\ell-1}_t K^{a,\ell-1}_t$.

$$K^{d,l}_t = -(R_d + B^{d,\ell-1^T}_t P^{d,l}_{t+1} B^{d,\ell-1}_t)^{-1} B^{d,\ell-1^T}_t P^{d,l}_{t+1} \hat{A}^{d,l}_t, \tag{9}$$

Table 1. Experiments parameters

Parameter	Description	Synthetic Network	Facebook Network
λ	Stubbornness	0.70	0.70
η	Learning rate	0.50	1
α	Sharing probability	0.30	0.5
σ	Homophily coefficient	1	0.32
$\bar{\ell}$	Maximum cognition level	10	10
n	Number of Individuals	3000	4038

where $P_t^{d,l}$ satisfies

$$P_t^{d,l} = Q_d + (\hat{A}_t^{d,l})^T \left(P_{t+1}^{d,l} - P_{t+1}^{d,l} B_t^{d,\ell-1} S_t^{d,l^{-1}} \left(B_t^{d,\ell-1} \right)^T P_{t+1}^{d,l} \right) \hat{A}_t^{d,l},$$

$$S_t^{d,l} = R_d + B_t^{d,\ell-1^T} P_{t+1}^{d,l} B_t^{d,\ell-1}.$$

The Adversary (*leader*) then consider the ℓ-th level response of the defender $u_t^{d,l}(s) = K_t^{d,l} \bar{s}_t^{\ell-1}$, and forms its optimization problem similar to (8). Adversary (*leader*) solves the optimization using standard Riccati recursion similar to the defender (*follower*) using the modified state matrix, $\hat{A}_t^{a,l} = A_t^{\ell-1} + B_t^{d,\ell-1} K_t^{d,\ell}$.

We then update the estimate of the trajectories, $\hat{s}_{t:t+H}^{\ell}$, according to the ℓ-th level feedback policies. We repeat this approach until we reach the maximal cognitive level assumed for each player. After the last step, we solve the defender problem one more time, to maintain the Stackelberg hierarchy, assuming that defender observes the adversary policy, and find the best response to that policy. The players then apply the policies over short span of time, and then get exact updates from the social network. This model-predictive implementation captures the empirical reality that reliable forecasts on social platforms rarely extend beyond a few days.

4 Results and Discussion

We investigate the effect of network connectivity on a synthetic network constructed according to homophily interactions and a real network induced by Facebook data [15]. Table 1 provides the parameters we used for each network. We utilized the *NVIDIA Ada Lovelace L4 Tensor Core GPU* for all experiments and the JAX package for automated derivatives to linearize dynamics and approximate equilibria. We assume a two-dimensional opinion space ($d = 2$) for both cases. For the real network, we used the Fruchterman-Reingold force-directed algorithm to find 2-D embeddings[1].

---

[1] All of the experiment codes used to generate the results in this section are accessible from https://github.com/ege-bayiz/influence-opinion-games.

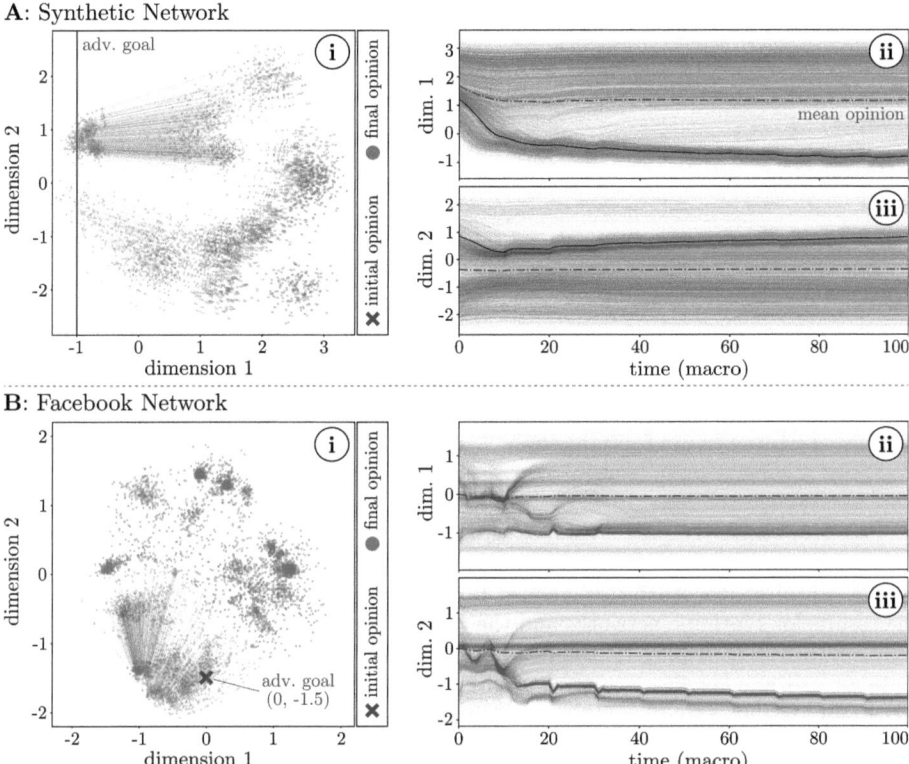

Fig. 1. Trajectories of individuals under competition. Figures **A** and **B** show the results for synthetic networks and the Facebook network, respectively. From left to right, both panels display (i) the scatter plot of how each individual's opinion changes between initial and final state, and (ii, iii) the opinion changes of all individuals across macro-time steps for both principal dimensions. The dashed red lines show the average opinion of the population across time.

Figure 1-(A,B) illustrates (i) the evolution of the network and (ii-iii) the individual trajectories of the synthetic and Facebook networks, respectively. For the synthetic network, the adversary's cost function considers only the first dimension; that is, the adversary aims solely to shift the first dimension of individual opinions toward -1. In the Facebook network, the adversary targets the opinion point $x_a = [\,0\,,\,-1.5\,]$, which corresponds to an adversarial cost matrix $Q_a = 3I$. In both scenarios, we set the adversary's regulation cost $R_a = 20I$. The defender has target and regulation costs of $Q_d = I$ and $R_d = 80I$, respectively, modeling the fact that taking extreme actions is substantially more costly for the defender than for the adversary.

We observe that in both scenarios (Fig. 1-(A,B)-i), the adversary targets nearby clusters and shifts their opinions toward its target. Especially in the synthetic network, the adversary manages to form its own echo chamber by

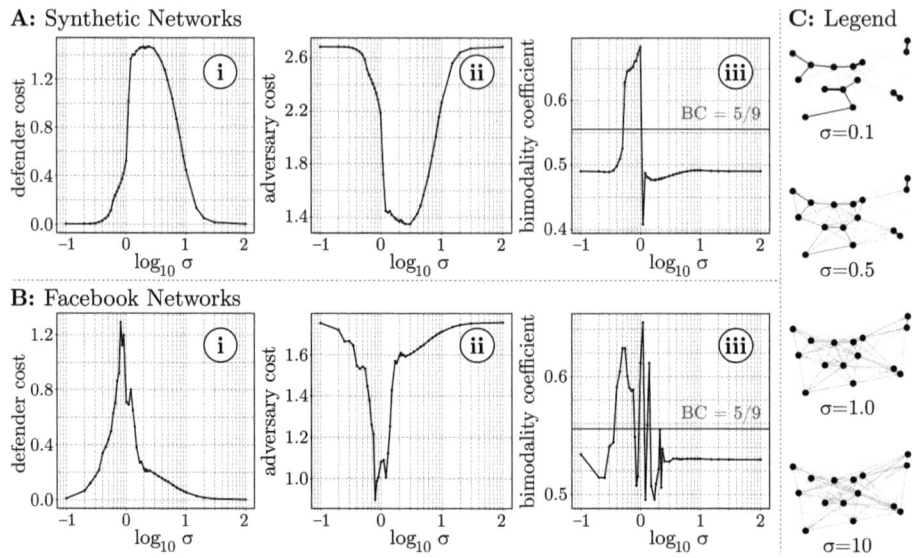

Fig. 2. Final state statistics across homophily kernels. Figures **A** and **B** show the results for synthetic networks and the Facebook network, respectively. From left to right, both panels display (i) the mean distance to the defender's goal, (ii) the mean distance to the adversary's goal, and (iii) the bimodality coefficient of the final state x_T. Figure **C** shows small sample networks for different kernel parameters, illustrating how the network topology varies with the kernel parameter σ.

exploiting the second dimension of the opinion space that it has freedom in (Fig. 1-A-iii). By aligning the second dimension with a dense cluster, the adversary gains control of a large portion of the population and manages to drive the opinions of a few clusters toward its objective(Figure 1-A-i). We observe in Fig. 1-(A,B)-ii that the equilibria result in an asymmetry among individuals; part of the network moves toward the adversary's target, while the defender manages to maintain the status quo for the remainder. This observation aligns with empirical results that show an asymmetric adverse influence [16,22]. Our results suggest that this asymmetry arises because the adversary exploits the underlying network topology and invests in specific parts of the network, rather than attempting to shift all individuals. Validating this observation and establishing a causal relation requires further theoretical and empirical experiments.

Figure 2-(A,B) presents the sensitivity analysis of the defender and adversary costs with respect to the homophily coefficient, σ. In both scenarios, we observe a similar trend; low and high values of σ make the network more resilient against the adversary. Low values of σ indicate strong local connections and weak global connections, as demonstrated in Fig. 2-C (for $\sigma = 0.1$). High locality hinders the adversary's ability to exploit network topology—low connectivity interrupts massage passing— making the formation of a strong adversarial community difficult. In contrast, for large σ values, the network approaches a complete graph with identical connections, as depicted in Fig. 2-C (for $\sigma = 10$). In this scenario, regardless of the locations the adversary and defender choose to start their messages, the network's symmetry cancels their effects, thereby preventing the adversary from shifting opinions.

However, social networks typically operate with moderate σ values; they are neither purely local, which would prevent the circulation of information, nor complete graphs, which would overwhelm users with unrelated information. Figure 2 shows that with moderate values of the homophily coefficient, the adversary gains an advantage and can shift a portion of the population towards its target. We emphasize that our approach approximates local linearized solutions; consequently, for the real network, we observe fluctuations in the costs of the defender and the adversary (Fig. 2-B-(i,ii)).

Traditional approaches to counter adversarial influence often focus on network control, such as by removing edges or altering the information flow [1,2]. However, our results demonstrate an important phenomenon. When dealing with a strategic adversary, understanding the current level of homophily is crucial, as adjusting it incorrectly can further exacerbate the problem and enable additional exploitation by the adversary.

Figure 2-(A,B)-iii demonstrates the bimodality coefficient, β, with respect to the homophily coefficient, σ. Our results reveal a phase transition for the synthetic network. As we increase σ, the bimodality coefficient initially increases, indicating that the adversary only captures a portion of the population. However, increasing σ beyond a certain threshold changes the equilibrium behavior. In this phase, the adversary cam attracts most of the population, creating a uni-modal distribution around its target. As σ increases further, the adversary gradually loses this advantage and the population shifts back to its initial state.

5 Conclusion

We study the effect of network topology to adversarial information operation in online social networks by introducing a novel bi-level influence-opinion game model. We formulate the problem as a Stackelberg game and present a scalable algorithm to approximate the equilibria for players with limited cognition. This approach captures correlation effects between the message propagation and the evolution of opinions alongside adaptive strategies that are often neglected in standard models. Our empirical investigations on synthetic and real-world Facebook networks revealed that network topology, particularly network connectivity, critically shapes adversarial resilience. We observe that both highly local and complete network topologies enhance network resilience, while moderate levels of connectivity can render networks vulnerable to adversarial influence, often leading to asymmetric opinion shifts, polarization, and favoring adversaries. These findings underscore the non-trivial impact of network topology and highlight that interventions aimed at altering network structure must be carefully considered, contributing to a deeper understanding of information conflict dynamics for designing more robust social platforms and effective counter-influence strategies.

Acknowledgments. We would like to thank Ashwin Ram for his discussions in the preliminary stages of problem formulation.

This work was supported in parts by ARO under grant number W911NF-23-1-0317, by DARPA under grant number HR001123S0001, and by ONR under grant number N00014-22-1-2703.

Disclosure of Interests. The authors have no competing interests to declare that are relevant to the content of this article.

References

1. Amini, A., Bayiz, Y.E., Topcu, U.: Control of misinformation with safety and engagement guarantees. In: 2024 American Control Conference (ACC), pp. 151–158. IEEE (2024)
2. Bayiz, Y.E., Topcu, U.: Optimization-based countering of misinformation on social networks. In: 2024 American Control Conference (ACC), pp. 135–142. IEEE (2024)
3. Behboodian, J.: On the modes of a mixture of two normal distributions. Technometrics **12**(1), 131–139 (1970). https://doi.org/10.1080/00401706.1970.10488640
4. Borodin, A., Braverman, M., Lucier, B., Oren, J.: Strategyproof mechanisms for competitive influence in networks. In: Proceedings of the 22nd international conference on World Wide Web, pp. 141–150 (2013)
5. Budak, C., Agrawal, D., El Abbadi, A.: Limiting the spread of misinformation in social networks. In: Proceedings of the 20th International Conference on World Wide Web. pp. 665–674 (2011)
6. Friedkin, N.E., Johnsen, E.C.: Social influence and opinions. J. Math. Sociol. **15**(3-4), 193–206 (1990)
7. Gaitonde, J., Kleinberg, J., Tardos, E.: Adversarial perturbations of opinion dynamics in networks. In: Proceedings of the 21st ACM Conference on Economics and Computation, pp. 471–472 (2020)
8. Golpayegani, F., et al.: Adaptation in edge computing: a review on design principles and research challenges. ACM Trans. Auton. Adapt. Syst. **19**(3), 1–43 (2024)
9. Gu, Y., Sun, Y., Gao, J.: The co-evolution model for social network evolving and opinion migration. In: Proceedings of the 23rd ACM SIGKDD International Conference on knowledge Discovery and Data Mining, pp. 175–184 (2017)
10. Hajaj, C., Joveski, Z., Yu, S., Vorobeychik, Y.: Robust coordination in adversarial social networks: from human behavior to agent-based modeling. Netw. Sci. **9**(3), 255–290 (2021)
11. Jia, P., MirTabatabaei, A., Friedkin, N.E., Bullo, F.: Opinion dynamics and the evolution of social power in influence networks. SIAM Rev. **57**(3), 367–397 (2015)
12. Kandel, D.B.: Homophily, selection, and socialization in adolescent friendships 1. In: Interpersonal development, pp. 249–258. Routledge (2017)
13. Kempe, D., Kleinberg, J., Tardos, É.: Maximizing the spread of influence through a social network. In: Proceedings of the ninth ACM SIGKDD International Conference on Knowledge Discovery and Data Mining, pp. 137–146 (2003)
14. Kurtz-David, V., Brandenburger, A., Glimcher, P.: The limits of social cognition: production functions and reasoning in strategic interactions. SSRN (2024)
15. Leskovec, J., Mcauley, J.: Learning to discover social circles in ego networks. Adv. Neural Info. Process. Syst. **25** (2012)

16. Rao, A., Morstatter, F., Lerman, K.: Partisan asymmetries in exposure to misinformation. Sci. Rep. **12**(1), 15671 (2022)
17. Rosas, A.: Evolutionary game theory meets social science: is there a unifying rule for human cooperation? J. Theor. Biol. **264**(2), 450–456 (2010)
18. Simaan, M., Cruz, J.B., Jr.: On the Stackelberg strategy in nonzero-sum games. J. Optim. Theory Appl. **11**(5), 533–555 (1973)
19. Stella, M., Ferrara, E., De Domenico, M.: Bots increase exposure to negative and inflammatory content in online social systems. Proc. Natl. Acad. Sci. **115**(49), 12435–12440 (2018)
20. Truong, B.T., Lou, X., Flammini, A., Menczer, F.: Quantifying the vulnerabilities of the online public square to adversarial manipulation tactics. PNAS Nexus **3**(7), 258 (2024)
21. Vamvoudakis, K.G., Fotiadis, F., Kanellopoulos, A., Kokolakis, N.M.T.: Nonequilibrium dynamical games: a control systems perspective. Annu. Rev. Control. **53**, 6–18 (2022)
22. Williamson, P.: Take the time and effort to correct misinformation. Nature **540**(7632), 171–171 (2016)
23. Zannettou, S., Caulfield, T., Bradlyn, B., De Cristofaro, E., Stringhini, G., Blackburn, J.: Characterizing the use of images in state-sponsored information warfare operations by Russian trolls on twitter. In: Proceedings of the International AAAI Conference on Web and Social Media, vol. 14, pp. 774–785 (2020)
24. Zuo, J., Liu, X., Joe-Wong, C., Lui, J.C., Chen, W.: Online competitive influence maximization. In: International Conference on Artificial Intelligence and Statistics, pp. 11472–11502. PMLR (2022)

The Power of Stories: Narrative Priming in Networked Multi-Agent LLM Interactions

Gerrit Großmann[1], Larisa Ivanova[1,3](✉), Sai Leela Poduru[1,2], Mohaddeseh Tabrizian[1,2], Islam Mesabah[1], David A. Selby[1], and Sebastian J. Vollmer[1,2]

[1] German Research Center for Artificial Intelligence (DFKI), Kaiserslautern, Germany
larisa.ivanova@dfki.de
[2] Department of Computer Science, University of Kaiserslautern–Landau (RPTU), Kaiserslautern, Germany
[3] Department of Language Science and Technology, Saarland University, Saarbrücken, Germany

Abstract. Research suggests that large-scale human cooperation is driven by shared narratives that encode common beliefs and values. This study explores whether such narratives can similarly nudge LLM agents toward collaboration.

Therefore, we let LLM agents play a (networked) finitely repeated public goods game after being primed with different stories.

Our experiments address four questions: (1) How do narratives influence negotiation behavior? (2) What differs when agents share the same story versus different ones? (3) What happens when the agent numbers grow? (4) Are agents resilient against self-serving participants?

We find that story-based priming significantly affects collaboration. *Common* stories improve collaboration and benefit all participants, while *different* story priming reverses this effect, favoring self-interested agents. These patterns persist across network sizes and structures.

These findings have implications for multi-agent coordination and AI alignment.

Code is available at github.com/storyagents25/story-agents.

Keywords: LLM Agents · Narrative Priming · Collaboration and Competition · Cooperation in Networks

1 Introduction

From ancient creation myths uniting scattered tribes to modern national narratives binding millions of strangers, shared stories are humanity's most powerful technology for large-scale cooperation [5,13,14]. These collective narratives enable coordinated actions across vast networks of strangers who would otherwise

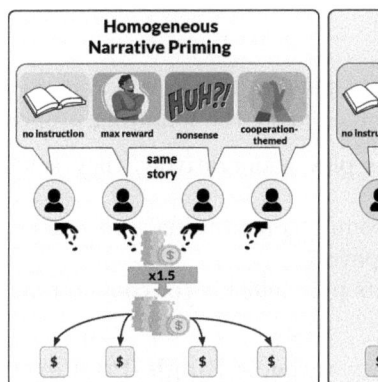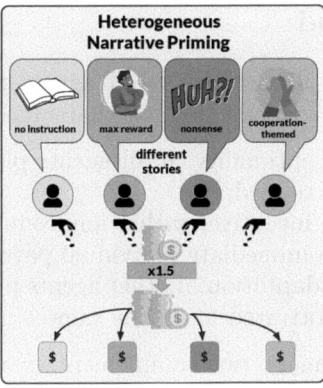

Fig. 1. Repeated public goods game with narrative priming. *Homogeneous*: all agents receive identical story prompts. *Heterogeneous*: each agent receives different narrative priming, creating mixed behavioral contexts within the same game.

lack mutual trust. As our world becomes increasingly populated by (multimodal) LLM agents deployed in complex multi-agent environments that require cooperation and competition [21,31,34], a critical question emerges: can we apply humanity's most successful cooperation mechanism to artificial agents?

While prior work has assessed LLM agents' cooperation [1,4] and LLM alignment [8,20], the potential for shared narratives to enhance collaboration in agent networks remains largely unexplored. Drawing inspiration from the role of stories in human cooperation, we investigate whether narrative priming can effectively promote collaborative behavior among LLM agents.

To test this, we use the Public Goods game, a framework which creates controlled conflicts between individual and collective benefit: agents receive endowments and decide whether to contribute personal resources to shared pools that multiply and redistribute equally regardless of individual input. This creates a classic cooperation dilemma where collective welfare demands full contribution, while individual rationality favors free-riding [3,28,32].

Our experiments test how narrative priming affects cooperation in networked environments where agents participate in overlapping groups, mirroring realistic interconnected social relationships and cooperation decisions [28,32]. We examine system-wide collaboration and individual outcomes among LLM agents. We also test the robustness of cooperation in the presence of selfish individuals.

Our results indicate that shared narratives improve cooperation when all participants receive identical cooperation-themed story prompts, but this effect reverses in heterogeneous groups with mixed narrative priming. These patterns persist even in networked settings with multiple pools, suggesting that narrative coherence *is* important for effective cooperation.

2 Method

Our method is based on LLM agents playing together a repeated networked game of public goods (Fig. 1), characterized by the following properties:

1. Collective optimality: if all agents play cooperatively, they achieve a higher individual reward;
2. Individual incentive: within any round, contributing zero tokens to any pool maximizes immediate individual payoff;
3. Iterative adaptation: if other agents play selfishly (or cooperatively), an agent may be motivated to do the same.

We implement two complementary variants: **Single-Pool** experiments use one shared pool to test scaling effects across group sizes and robustness to defection, while **Multi-Pool** experiment introduces strategic complexity through overlapping pools.

In both paradigms, we manipulate behavioral homogeneity through *narrative priming* (see Sect. 2.2) to examine whether story-based conditioning affects agents' cooperative strategies.

2.1 Game Procedure

In each game, we instantiate N agents with game rules, assigned narratives, and assigned pools. Each game consists of R rounds, with each round r following a fixed sequence:

1. *Endowment*: each agent i receives T tokens;
2. *Contribution*: for each of their M assigned pools p, agents decide contribution amounts t ($t \in \mathbb{Z}$ with $0 \leq t \leq T$);
3. *Payoff calculation*: payoffs are calculated (see subsequent paragraph) and redistributed among pool members;
4. *Feedback*: agents receive complete information about all relevant contributions and payoff breakdowns.

Data Collection and Metrics. We collect individual agent contributions per round (per pool in multi-pool experiment), round payoffs, and cumulative payoffs across all rounds within each game. Primary metrics are cumulative payoffs per agent (Eq. (1)), collaboration scores (Eq. (2)), and, in multi-pool experiment, global vs local pool preference ratios.

Payoff Calculation. Each pool p has member set M_p and collects total contributions $T_p = \sum_{i \in M_p} t_{i,p}$. These contributions are multiplied by a fixed factor m and redistributed equally among pool members. Agent i's total round payoff π_i consists of two components: the agent's share of the returns from all pools it participated in, and the unspent remainder of its initial endowment T:

$$\pi_i = \sum_{p:\, i \in M_p} \frac{m T_p}{|M_p|} + \left(T - \sum_p t_{i,p}\right). \tag{1}$$

Table 1. Narrative prompt properties: *lexical diversity* (vocabulary richness, from low/repetitive (0) to high/diverse (1)), *sentiment score* (emotional valence, from negative (−1) to positive (+1)).

Story Type	Prompt	Token Count	Lexical Diversity	Sentiment Score	Main Theme	Cultural Origin
Baseline	noinstruct	0	0.000	N/A	N/A	N/A
	maxreward	10	0.889	0.300	None	N/A
	nsCarrot	320	0.596	0.150	Curiosity & self-reward	Invented modern fantasy
	nsPlumber	305	0.560	0.287	Creative problem solving	Invented modern fantasy
Meaningful	OldManSons	220	0.636	-0.042	Strength through (familial) unity	European folktale
	Odyssey	322	0.677	0.150	Resilience via wisdom & alliances	Classical Greek epic
	Soup	285	0.597	0.075	Resource pooling & generosity	European folktale
	Peacemaker	256	0.640	0.404	Unity through dialogue & consensus	Iroquois (Indigenous legend)
	Musketeers	273	0.643	0.330	Unity through strategic alliance	French adventure novel
	Teamwork	309	0.598	0.060	Combining strengths to succeed	Modern illustrative parable
	Spoons	867	0.402	0.145	Mutual aid via sharing	European allegory
	Turnip	324	0.595	0.085	Every contribution matters	European folktale

Collaboration Score. We measure cooperation effectiveness as the proportion of total possible contributions actually made across all agents and rounds in a game:

$$\text{Collaboration Score} = \frac{\sum_{r=1}^{R} \sum_{i=1}^{N} t_{i,r}}{N\,R\,T}, \quad (2)$$

where $t_{i,r}$ is agent i's contribution in round r. A score of 1.0 indicates perfect cooperation, while lower values reflect deviations due to reduced participation or strategic choices. This metric serves as a key proxy for evaluating how effectively different narratives influence agent cooperation.

2.2 Narrative Priming

To test the effect of narrative priming on agents' collaborative behavior, we prime them with story-based behavioral context via system prompt:

> "Your behavior is influenced by the following bedtime story your mother read to you every night: [Story]"

Our story corpus (cf. Table 1, also available on GitHub) comprises 8 cooperation-themed narratives emphasizing teamwork and collective benefit, plus 4 control conditions including no instructions, explicit self-interest directives, and two nonsensical stories lacking coherent themes. Stories were selected to balance cultural diversity and summarized to retain core cooperation-themed elements while minimizing extraneous narrative details. Depending on experimental condition, agents receive either identical stories (homogeneous condition) or randomly sampled distinct stories (heterogeneous condition) from this corpus.

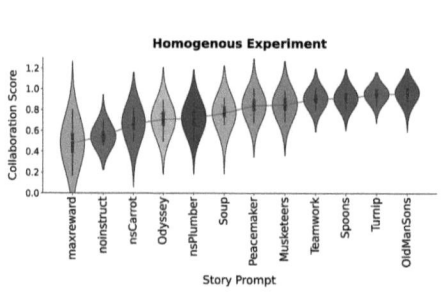 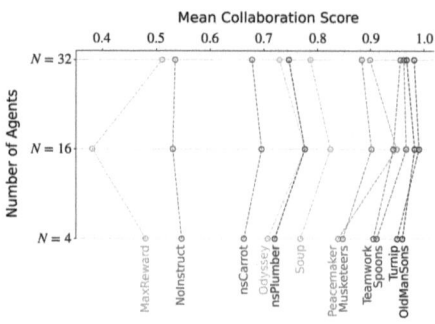

(a) Collaboration scores for homogeneous group ($N = 4$). Pink-shaded cooperation-themed stories get higher collaboration scores than blue-shaded controls. Gray trend line represents the mean.

(b) Scaling experiment results for homogeneous agents across different group sizes. The ranking remains relatively consistent as group size increases.

Fig. 2. Narrative priming effects on cooperation in homogeneous groups.

3 Results

Implementation. We used *Meta-Llama-3.1-70B-Instruct-FP8*[1], and conducted experiments using varied temperature parameters (0.6, 0.8, 1.0). At higher temperatures, the priming effects show stronger differentiation, and negotiation dynamics become less pronounced. For clarity and consistency, our detailed analysis focuses on experiments run with temp = 0.6.

3.1 Single-Pool Cooperation

The Single-Pool experiments use variable group sizes ($N \in \{4, 16, 32\}$) playing $R = 5$ rounds with one shared pool, running 100 games per story per group size, $m = 1.5$.

Exp. 1.1: Cooperation Among Homogeneous Agents. In 4-agent groups, cooperation-themed stories ("OldManSons," "Turnip") achieve near-perfect collaboration scores, significantly outperforming baseline controls. Self-interest ("maxreward") and nonsensical narratives yield noticeably lower scores (Fig. 2a). These findings suggest that narrative priming has a measurable effect on reinforcing cooperative behavior in multi-agent systems.

Exp. 1.2: Scaling Effects. Cooperation patterns remain consistent across network sizes $N \in \{4, 16, 32\}$, with relative narrative rankings preserved. Larger agent networks exhibit more pronounced differences between cooperative and baseline conditions (Fig. 2b).

---

[1] huggingface.co/neuralmagic/Meta-Llama-3.1-70B-Instruct-FP8.

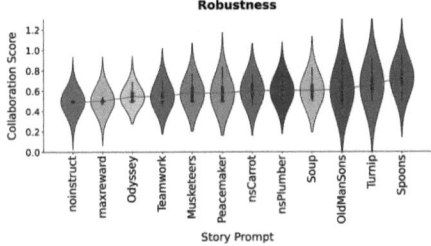 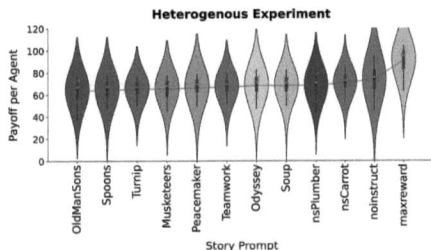

(a) Robustness Testing: one free-rider in a group of four. Overall cooperation decreases compared to the baseline.

(b) Cumulative payoffs in mixed narrative groups ($N = 4$), aggregated by story prompt. Self-interested agents achieve highest returns.

Fig. 3. Cooperation under disruption: narrative-based adaptation to free-riders and breakdown under narrative misalignment.

Exp. 1.3: Robustness Testing. To assess strategic adaptation when confronted with exploitative agents, we introduced persistent free-riders (always contributing zero) in 4-agent groups. Results reveal that agents dynamically adapt their strategies based on environmental (narrative) context rather than using fixed contribution patterns (Fig. 3a). This adaptive adjustment functions as implicit negotiation: agents recalibrate contributions in response to others' behavior, balancing cooperation and self-interest.

Exp. 1.4: Heterogeneous Agents. Mixed narrative conditions ($N = 4$, 400 games) reversed cooperation dynamics. Self-interested agents ("maxreward") achieved highest cumulative payoffs (90.87 ± 10.06), while cooperation-primed agents ("OldManSons," "Spoons") obtained lowest returns (Fig. 3b). This inversion demonstrates that narrative coherence among playing agents does determine the viability of cooperation.

3.2 Multi-Pool Cooperation

Real-world cooperation and resource allocation often occur across overlapping contexts [23,30]. $N = 4$ agents play $R = 10$ rounds across $M = 3$ overlapping pools: one global pool (all 4 agents) and two smaller pools (2 agents each, randomly assigned), running 10 games per story for homogeneous and 100 games for heterogeneous conditions, $m = 1.5$. Each agent belongs to exactly two pools, requiring strategic resource allocation across competing collective interests. Agents are primed with the same cooperation-themed or baseline narratives as in single-pool experiments.

Exp. 2: Resource Allocation Dynamics. Narrative priming effects observed in single-pool experiments persist: under homogeneous priming, cooperation-themed stories achieve higher collaboration scores (Fig. 4a) and preferentially allocate most of their tokens to global pools (Fig. 4b). However, mixed

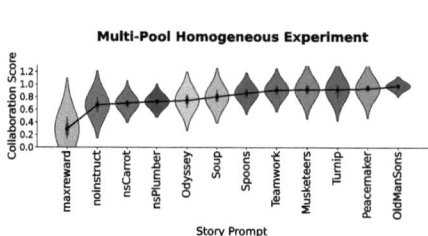
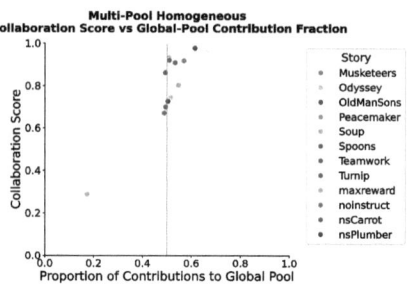
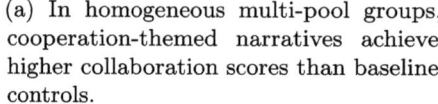

(a) In homogeneous multi-pool groups, cooperation-themed narratives achieve higher collaboration scores than baseline controls.

(b) Story-level collaboration scores vs global pool allocation fractions reveal cooperation-themed narratives achieve higher scores through greater global pool contributions.

Fig. 4. Multi-pool homogeneous condition: cooperation-themed narratives achieve higher collaboration scores and payoffs through preferential global pool allocation.

story priming (heterogeneous) again reverse outcomes, with self-interest agents ("maxreward") achieving highest payoffs (Fig. 6a in Appendix), while agents primed with cooperation-themed stories achieved lower payoffs (Fig. 6b in Appendix).

4 Discussion

Our experiments show that narrative priming systematically affects how LLM agents collaborate and compete across repeated networked public goods games. These dynamics reflect implicit negotiation, where strategies shift round-by-round without explicit dialogue. The consistency of effects (enhanced cooperation under shared narratives, competitive dynamics under mixed priming) across network topologies suggests robust narrative-driven cooperation mechanisms.

However, the interpretation of these results remains open. If the goal is simply to induce a specific strategy, one could simply prompt agents with direct instructions. The more intriguing question concerns how implicit or adversarial priming leads to unintended behavioral strategies.

The causal mechanisms underlying this phenomenon are still unclear. Notably, narratives that encourage collaboration contain teamwork-related vocabulary even at a statistical (bag-of-words) level making it difficult to isolate narrative structure from semantic content. Preliminary results suggest that cooperation-themed stories analyzed purely at the lexical level (removing narrative structure and context) still produce cooperative strategies though the effect appears weaker than when full narrative context is preserved. These findings require more rigorous validation to determine whether semantic content alone drives the observed behaviors or whether narrative coherence provides additional cooperative influence. It would be interesting to explore whether subtler narratives produce similar effects. Additionally, these narratives may resemble text from the training corpus and activate related contexts during inference. Preliminary results indicate that narratives emphasizing self-care over teamwork yield strategies comparable to those observed under the "maxreward" prompt. The role of Reinforcement Learning from Human Feedback (RLHF), a key component of LLM training, in shaping this behavior also remains uncertain. Furthermore, the selected stories were not rigorously controlled for emotional valence or complexity, which could confound the results.

The systematic reversal under heterogeneous priming reveals a basic coordination breakdown: when agents don't share the same behavioral cues, self-interested strategies consistently exploit cooperative ones, but the precise mechanisms require closer examination through controlled studies varying degrees of narrative alignment and measuring intermediate coordination signals. The consistent effects across group sizes challenge standard assumptions about cooperation breaking down in larger groups [25], though this apparent scale-invariance should be validated with larger agent populations and diverse network typologies to determine the boundaries of narrative-based coordination.

The progression from single-pool to multi-pool networks reveals that narrative coherence becomes increasingly critical as network complexity grows. In overlapping pool structures, mixed narratives create strategic conflicts that consistently favor individually rational agents, while shared narratives enable coordination across multiple resource domains simultaneously.

Overall, we do not interpret these experiments as evidence of human-like priming in LLMs. There is also a risk of anthropomorphizing the model's behavior—while agents may appear cooperative, their responses are likely driven by statistical patterns in the training data rather than deliberate reasoning.

5 Related Work

Prior work in game theory and economics demonstrates that cooperation is influenced by factors such as communication [2,33], shared norms [28,32], and strategic alignment [12,19]. Psychological research demonstrates that priming can affect social behavior [16,24]. While LLM multi-agent systems display various social dynamics [21,26], there is limited exploration of narrative-driven priming, analogous to cultural storytelling, and its impacts on these dynamics [7,15]. Our work aims to fill this gap by testing whether shared narratives serve as "cultural glue" similar to prosocial norms in experimental economics, focusing on how narrative context shapes cooperation in repeated public goods games.

Collaboration Conceptualizations. Economic games model collaboration vs. competition trade-offs, highlighting individual versus collective interests [28,32]. Evolutionary models showcase how moderate cooperation emerges from coevolution of behavior [29], while excessive greed can destabilize societies [29]. Empirical studies indicate decline in contributions over rounds, suggesting that multi-round dynamics create opportunities for fostering reciprocity and conditional cooperation [11], which parallels how narrative priming might influence outcomes in multi-agent systems.

Negotiations. In game theory and economics, negotiation frameworks emphasize strategic reasoning, value creation, and rational decision-making, encompassing various frameworks and strategies [6,17,33]. Repeated Prisoner's Dilemma studies show how strategic uncertainty shapes cooperation [9,18,27]. Our study introduces narrative priming as a variable that reshapes agents' perceived priorities, thereby extending classical models to account for story-driven shifts in cooperation behavior in networked resource allocation.

LLM Sociology and Multi-Agent Collaboration. Recent studies position LLMs as proxies for studying human-like social dynamics, replicating behaviors in strategic games [1,2,10] and multi-agent systems. Despite these advancements, LLMs struggle with nuanced strategies like preference inference [4,19]. Integrative frameworks link game theory with collaborative workflows [21,31,34], aligning with "Cooperative AI" visions for bridging AI and social sciences [8,26].

Psychology and Priming. Psychological evidence indicates that priming influences cooperation through shared identities [35], prosocial modeling [16], and moral framing [24]. Exposure to stories boosts theory-of-mind skills [22], and structural priming in LLMs [15] suggests that narrative techniques can effectively guide LLM behavior.

6 Conclusions and Future Work

This study identifies narrative priming as a potential lever for steering collaboration in multi-agent systems: *common* stories improve cooperation across network topologies while *different* narratives favor competitive strategies, with effects persisting across single- and to multi-pool architectures.

Future work must examine causal mechanisms (e.g., via mechanistic interpretability) to trace how narrative inputs alter attention patterns or value representations in transformer layers. Temporal studies should evaluate whether priming effects decay over repeated games, while adversarial narratives should assess whether priming with malicious narratives destabilizes multi-agent systems. Additionally, cross-genre experiments (e.g., deception-focused stories) and scaling laws for agent populations will help map the semantic and structural boundaries of narrative priming. Comparative cross-model analysis (smaller architectures, non-RLHF variants) will be essential. Future work should also systematically examine narrative structure, emotional valence and varying degrees of cooperativeness.

Finally, a promising direction for future work is to empirically analyze the strategies of LLMs under different narrative primings and map these to empirical human strategies or theoretical results.

Ethical Considerations. A key concern surrounding LLMs is their environmental impact due to high computational requirements. We used Llama 3.1 (70B) on GH200 GPU with 1.4 kW power consumption, totaling 57.4 kWh over 41 h for 370,400 model calls processing 1.2B tokens.

Full paper (with Appendix) available at GitHub: Networked_Story_Agents.pdf.

Disclosure of Interests. The authors declare no competing interests.

References

1. Abdelnabi, S., Gomaa, A., Sivaprasad, S., Schönherr, L., Fritz, M.: Cooperation, competition, and maliciousness: LLM-stakeholders interactive negotiation. In: Advances in Neural Information Processing Systems, vol. 37, pp. 83548–83599 (2024)
2. Aher, G., Arriaga, R.I., Kalai, A.T.: Using large language models to simulate multiple humans and replicate human subject studies. In: Proceedings of the 40th International Conference on Machine Learning. ICML'23 (2023)
3. Andreoni, J.: Impure altruism and donations to public goods: a theory of warm-glow giving. Econ. J. **100**(401), 464–477 (1990)
4. Bianchi, F., Chia, P.J., Yuksekgonul, M., Tagliabue, J., Jurafsky, D., Zou, J.: How well can LLMs negotiate? NegotiationArena platform and analysis. In: Proceedings of the 41st International Conference on Machine Learning (2024)
5. Boyd, R., Richerson, P.J.: Culture and the evolution of human cooperation. Philosoph. Trans. Royal Soc. B Biol. Sci. **364**(1533), 3281–3288 (2009)
6. Brams, S., Quarles, R.J., McElreath, D.H., Waldron, M.E., Milstein, D.E.: Negotiation Games. Routledge (2002)
7. Bullock, O.M., Shulman, H.C., Huskey, R.: Narratives are persuasive because they are easier to understand: examining processing fluency as a mechanism of narrative persuasion. Front. Commun. **6**, 719615 (2021)
8. Dafoe, A., Bachrach, Y., Hadfield, G., Horvitz, E., Larson, K., Graepel, T.: Cooperative AI: machines must learn to find common ground. Nature **593**(7857), 33–36 (2021)
9. Embrey, M., Fréchette, G.R., Yuksel, S.: Cooperation in the finitely repeated prisoner's dilemma. Q. J. Econ. **133**(1), 509–551 (2018)
10. (FAIR) Meta Fundamental AI Research Diplomacy Team (FAIR), et al.: Human-level play in the game of diplomacy by combining language models with strategic reasoning. Science **378**(6624), 1067–1074 (2022)
11. Fischbacher, U., Gächter, S., Fehr, E.: Are people conditionally cooperative? Evidence from a public goods experiment. Econ. Lett. **71**(3), 397–404 (2001)
12. Gemp, I., et al.: Steering language models with game-theoretic solvers. In: Agentic Markets Workshop at ICML 2024 (2024)
13. Harari, Y.N.: Sapiens: A Brief History of Humankind. Random House (2014)
14. Henrich, J.: Cultural group selection, coevolutionary processes and large-scale cooperation. J. Econ. Behav. Organ. **53**(1), 3–35 (2004)
15. Jumelet, J., Zuidema, W., Sinclair, A.: Do language models exhibit human-like structural priming effects? In: Ku, L.W., Martins, A., Srikumar, V. (eds.) Findings of the Association for Computational Linguistics: ACL 2024, pp. 14727–14742. Association for Computational Linguistics (2024)

16. Jung, H., Seo, E., Han, E., Henderson, M.D., Patall, E.A.: Prosocial modeling: a meta-analytic review and synthesis. Psychol. Bull. **146**(8), 635 (2020)
17. Kıbrıs, Ö.: Cooperative game theory approaches to negotiation. In: Handbook of Group Decision and Negotiation, pp. 151–166 (2010)
18. Kreps, D.M., Milgrom, P., Roberts, J., Wilson, R.: Rational cooperation in the finitely repeated prisoner's dilemma. J. Econ. Theory **27**(2), 245–252 (1982)
19. Kwon, D., Weiss, E., Kulshrestha, T., Chawla, K., Lucas, G., Gratch, J.: Are LLMs effective negotiators? Systematic evaluation of the multifaceted capabilities of LLMs in negotiation dialogues. In: Findings of the Association for Computational Linguistics: EMNLP 2024, pp. 5391–5413. Association for Computational Linguistics
20. Li, H., et al.: Theory of mind for multi-agent collaboration via large language models. In: Proceedings of the 2023 Conference on Empirical Methods in Natural Language Processing, pp. 180–192. Association for Computational Linguistics
21. Li, X., Wang, S., Zeng, S., Wu, Y., Yang, Y.: A survey on LLM-based multi-agent systems: workflow, infrastructure, and challenges. Vicinagearth **1**(1), 9 (2024)
22. Mak, H.W., Fancourt, D.: Reading for pleasure in childhood and adolescent healthy behaviours: longitudinal associations using the millennium cohort study. Prev. Med. **130**, 105889 (2020)
23. Menczer, F., Fortunato, S., Davis, C.A.: A First Course in Network Science. Cambridge University Press (2020)
24. Mieth, L., Buchner, A., Bell, R.: Moral labels increase cooperation and costly punishment in a prisoner's dilemma game with punishment option. Sci. Rep. **11**(1), 10221 (2021)
25. Olson Jr, M.: The Logic of Collective Action: Public Goods and the Theory of Groups, vol. 124. Harvard University Press (1971)
26. Park, J.S., O'Brien, J., Cai, C.J., Morris, M.R., Liang, P., Bernstein, M.S.: Generative agents: interactive simulacra of human behavior. In: Proceedings of the 36th Annual ACM Symposium on User Interface Software and Technology (2023)
27. Raihani, N.J., Bshary, R.: Resolving the iterated prisoner's dilemma: theory and reality. J. Evol. Biol. **24**(8), 1628–1639 (2011)
28. Raub, W., Buskens, V., Corten, R.: Social Dilemmas and Cooperation. Handbuch Modellbildung und Simulation in den Sozialwissenschaften, pp. 597–626 (2015)
29. Roca, C.P., Helbing, D.: Emergence of social cohesion in a model society of greedy, mobile individuals. Proc. Natl. Acad. Sci. **108**(28), 11370–11374 (2011)
30. Siegenfeld, A.F., Bar-Yam, Y.: An introduction to complex systems science and its applications. Complexity **2020**(1), 6105872 (2020)
31. Sun, H., Wu, Y., Cheng, Y., Chu, X.: Game theory meets large language models: a systematic survey. arXiv preprint arXiv:2502.09053 (2025)
32. Thielmann, I., Böhm, R., Ott, M., Hilbig, B.E.: Economic games: an introduction and guide for research. Collabra: Psychology **7**(1), 19004 (2021)
33. Thompson, L.L., Wang, J., Gunia, B.C.: Negotiation. Group processes, pp. 55–84 (2012)
34. Tran, K.T., Dao, D., Nguyen, M.D., Pham, Q.V., O'Sullivan, B., Nguyen, H.D.: Multi-agent collaboration mechanisms: a survey of LLMs. arXiv preprint arXiv:2501.06322 (2025)
35. Van Lange, P.A., Rand, D.G.: Human cooperation and the crises of climate change, COVID-19, and misinformation. Annu. Rev. Psychol. **73**(1), 379–402 (2022)

Moment Constrained Optimal Transport for Energy Demand Management of Heterogeneous Loads

Julien Cardinal[1,2](✉), Thomas Le Corre[1,2], and Ana Bušić[1,2]

[1] Inria, Paris, France
julien.cardinal@inria.fr
[2] DI ENS, ENS, PSL University, Paris, France

Abstract. This paper addresses the problem of coordinating a large population of heterogeneous electrical loads, such as electric vehicles (EVs) and water heaters (WHs), under global operational constraints. We extend the Moment Constrained Optimal Transport for Control (MCOT-C) framework to accommodate multiple classes of agents with distinct dynamics and cost structures. Our formulation relies on a mean-field limit that captures agent heterogeneity through class-specific distributions. We propose a scalable gradient descent algorithm and a Model Predictive Control (MPC) scheme that enables online adaptation of this algorithm to uncertain or progressively revealed agent information. The proposed approach is validated through numerical experiments on real datasets [8,9] for EVs and WHs, demonstrating the effectiveness of this method in enforcing global constraints while preserving agent-level dynamics.

Keywords: Mean Field Control · Optimal Transport · Smart Grids

1 Introduction

Demand-side management (DSM) has emerged as a key strategy for enhancing the flexibility and reliability of modern power systems, particularly in the context of increasing penetration of renewable energy sources.[1] Unlike traditional generation-side methods, DSM enables system operators to influence and coordinate the consumption patterns of end-users in real-time or near-real time. This is achieved by leveraging the flexibility of a wide range of distributed energy resources (DERs). The aggregated control of such devices can provide valuable services to the grid, including load shifting, peak shaving, frequency regulation, and congestion management, thus contributing to system stability and economic

---

This work was carried out in the framework of the AI-NRGY project, funded by France 2030 (Grant No: ANR-22-PETA-0004).

[1] The code used to generate the results of this paper is available at https://github.com/Kreyparion/Heterogeneous-MCOT.

efficiency [5,6]. In demand-side management applications, the number of controllable devices is typically very large, which naturally motivates the use of mean-field control (MFC) approaches [1].

MFC problems are well-studied because of their versatility in domains ranging from economics and energy management to large-scale networked infrastructures [1,2]. When a large population of agents interacts through a common cost, the difficulty of solving the associated control problem increases with the number of agents. The mean-field approximation mitigates this difficulty by representing the system through a single, representative agent influenced by the aggregate state of the population, leading to decentralized control strategies that align individual behavior with a shared global objective.

In demand-side management control problems, the populations of devices are highly heterogeneous, both in their physical nature (such as electric vehicles, refrigerators, or water heaters), and in their individual characteristics (for instance differing battery capacities or charging rates in the case of electric vehicles). Blindly applying homogeneous decentralized strategies may result in suboptimal or even infeasible behavior [3]. To efficiently control heterogeneous agents, structured population models [1] or heterogeneous mean-field games [3] have been proposed. The challenge of controlling distributed energy resources has been addressed in [4], where a duality-based approach is employed, closely related to the one considered in this work.

Optimal transport (OT) provides an intuitive approach for modeling the evolution of the aggregate distributions considered in MFC. Moment-Constrained Optimal Transport (MCOT) has been proposed as a relaxation of the classical OT problem, where the marginal constraints are relaxed using moment inequalities [7]. This relaxation was first introduced to approximate the optimal transport problem but was recently considered as a formulation of a mean field control problem, for a homogeneous population [11]. In this work, we extend the MCOT framework to incorporate agent heterogeneity, by formulating a heterogeneous moment-constrained optimal transport problem.

Contributions: We develop a tractable optimization framework for the real-time coordination of heterogeneous loads, combining ideas from optimal transport and mean field control. The first contribution of this paper is to extend the framework of Moment Constrained Optimal Transport for Control (MCOT-C) [11] to a heterogeneous setting. The second contribution is to develop a Model Predictive Control (MPC) approach where the agents' data is progressively discovered during the day. The relevance of this method is checked on a real data set of Water Heaters and Electric Vehicles.

Organisation of the Paper: Section 2 introduces the formulation of the problem of controlling heterogeneous agents as a Moment Constrained Optimal Transport problem. The main results on the existence of solutions and expression of the dual and its gradients are given in this section, leading to a gradient descent algorithm. In Sect. 3, we propose an MPC version in which the data is discovered sequentially. An application illustrated this method for two types of agents: Electric Vehicles (dataset [8]) and Water Heaters (dataset [9]).

2 Heterogeneous MCOT

2.1 Heterogenous Control Problem

We consider a population of N agents, denoted $\{X^i\}_{1 \leq i \leq N}$, partitioned into H classes. Each class $h \in \{1, \ldots, H\}$ includes a subset $\mathcal{I}^{(h)} \subseteq \{1, \ldots, N\}$ of agents sharing common dynamics and cost. Let $N^{(h)} = |\mathcal{I}^{(h)}|$ denote the number of agents in class h, with $\sum_{h=1}^{H} N^{(h)} = N$. The proportion of each class is denoted $\alpha^{(h)} \colon \forall h \in \{1, \ldots, H\}, \ \alpha^{(h)} = \frac{N^{(h)}}{N}$.

A central agent coordinates the agents to satisfy a set of A global constraints:

$$\forall a \in \{1, \ldots, A\}, \quad \sum_{h=1}^{H} \sum_{i \in \mathcal{I}^{(h)}} f_a^{(h)}(X^i) \leq 0, \tag{1}$$

where $f^{(h)} \colon \mathcal{X}^{(h)} \to \mathcal{R}^A$ is a function depending on the class h.

An example of such a constraint is a total power consumption cap $P_{\max}$ that must not be exceeded at any time. Equality constraints, such as signal tracking, can be formulated as pairs of inequality constraints and are treated similarly, as in Sect. 3. Each agent is defined by their state $\{s^i\} \in \mathcal{S}$ and makes a control decision $\{w^i\} \in \mathcal{W}$, which induces a cost $c^{(h)}$ (examples of this cost are provided in Sect. 3). The central agent aims at minimizing the following problem:

$$\min_{\{w^i\}_{1 \leq i \leq N}} \left\{ \sum_{h=1}^{H} \sum_{i \in \mathcal{I}^{(h)}} c^{(h)}(s^i, w^i) : \sum_{i=1}^{N} f(s^i, w^i) \leq 0 \right\} \tag{2}$$

This formulation sets the stage for a class-wise mean field approach to control design, developed in the following sections.

2.2 Heterogeneous MCOT

We introduce in this section the Mean Field formulation [10] of Problem 2 to the case of H distinct classes of agents. Each class $h \in \{1, \ldots, H\}$ is associated with a population of agents sharing common nominal dynamics and costs. The behavior of class-h agents is described by a common distribution $\mu_1^{(h)} \in \mathcal{B}(\mathcal{X}^{(h)})$, where $\mathcal{B}(\mathcal{X}^{(h)})$ denotes the Borel probability measures on $\mathcal{X}^{(h)} = \mathcal{S}^{(h)} \times \mathcal{W}^{(h)}$, the product space of the state and control spaces. The aggregate system is thus described by the collection $\{\mu_1^{(h)}\}_{h=1}^{H}$, each representing a homogeneous subpopulation. The goal is to control each class by modifying its distribution, resulting in a transport of mass between distributions over $\mathcal{X}^{(h)}$.

We aim to identify joint distributions $\pi^{(h)} \in \mathcal{B}(\mathcal{X}^{(h)} \times \mathcal{X}^{(h)})$ for each class h, representing the coupling between initial and controlled trajectories. Each $\pi^{(h)}$ is subject to constraints reflecting both the nominal dynamics and control objectives specific to class h and is associated with a cost: $\langle \pi^{(h)}, c^{(h)} \rangle = \int c^{(h)}(x, y) \pi^{(h)}(x, y) \, dx dy$. We define the marginals $\pi_1^{(h)} = \int \pi^{(h)}(x, y) dy$ and

$\pi_2^{(h)} = \int \pi^{(h)}(x,y)dx$, with $\pi_1^{(h)} = \mu_1^{(h)}$ fixed. Second marginals $\pi_2^{(h)}$ must belong to the following set of distributions satisfying the global moment constraints:

$$\mathcal{P}_f = \left\{ \{\mu^{(h)}\}_{h=1}^H \in \prod_{h=1}^H \mathcal{B}(\mathcal{X}^{(h)}) : \forall a \in \{1,\ldots,A\}, \sum_{h=1}^H \alpha^{(h)} \langle \mu^{(h)}, f_a^{(h)} \rangle \leq 0 \right\} \quad (3)$$

We define the feasible coupling set, as bivariate distributions having for first marginal $\mu_1^{(h)}$ and whose two marginals coincide on the state space $\mathcal{S}$:

$$\mathcal{U}(\mu_1^{(h)}) = \left\{ \pi^{(h)} \in \mathcal{B}(\mathcal{X}^{(h)} \times \mathcal{X}^{(h)}) : \pi_1^{(h)} = \mu_1^{(h)} \right.$$
$$\left. \text{and } \forall s_1 \neq s_2 \in \mathcal{S}, \pi^{(h)}((s_1,.),(s_2,.)) = 0 \right\} \quad (4)$$

An entropic regularizer is introduced for computational reasons and to impose hard constraints on agent behavior [11]:

$$D_{\mathrm{KL}}(\pi^{(h)} \| \mu_1^{(h)} \otimes \mu_2^{(h)}) = \int \log\left(\frac{\pi^{(h)}(x,y)}{\mu_1^{(h)}(x)\mu_2^{(h)}(y)} \right) \pi^{(h)}(x,y)\, dx dy \quad (5)$$

leading to the formulation of the problem: **Multi-class MCOT-C**

$$\min_{\{\pi^{(h)}\}_{h=1}^H} \sum_{h=1}^H \alpha^{(h)} \left(\langle \pi^{(h)}, c^{(h)} \rangle + \varepsilon D_{\mathrm{KL}}(\pi^{(h)} \| \mu_1^{(h)} \otimes \mu_2^{(h)}) \right) \quad (6)$$
$$\text{s.t. } \forall h,\ \pi^{(h)} \in \mathcal{U}(\mu_1^{(h)}),\ \{\pi_2^{(h)}\}_{h=1}^H \in \mathcal{P}_f$$

2.3 Dual Problem

In this subsection, we begin by stating the assumptions, then present the dual formulation of the heterogeneous MCOT problem 6. We then show that this dual problem admits a closed-form expression (Proposition 1), which is differentiable, and we provide the explicit formula for its gradient (Proposition 2).

Assumptions: In the following, we make the following assumptions,

(A1) Each $c^{(h)}: \mathcal{X}^{(h)} \times \mathcal{X}^{(h)} \to \mathbb{R}_+$ and $f^{(h)}: \mathcal{X}^{(h)} \to \mathbb{R}^{A^{(h)}}$ are continuous. For each class h, $\mathcal{P}_{f,r}^{(h)} = \left\{ \mu \in \mathcal{B}(\mathcal{X}^{(h)}) : \forall a \in \{1,\ldots,A\}, \sum_{h=1}^H \langle \mu, f_a^{(h)} \rangle \leq r \right\}$ is non-empty for all r in a neighborhood of 0.
(A2) Each $\mu_1^{(h)}, \mu_2^{(h)}$ has compact support. The feasibility property under perturbations holds for each class.
(A3) The covariance $\Sigma^{(h)}$ of $Y^{(h)}$ is positive definite for $Y^{(h)} \sim \mu_2^{(h)}$.

Dual Problem: We denote the Lagrange multiplier associated to the constraints 1 by λ and we derive the dual problem from 6: $\forall \lambda \in \mathbb{R}^A$,

$$\varphi(\lambda) = \min_{\{\pi^{(h)}\}_{h=1}^H} \sum_{h=1}^H \alpha^{(h)} \left(\varepsilon D_{\mathrm{KL}}(\pi^{(h)} \| \mu_1^{(h)} \otimes \mu_2^{(h)}) - \langle \pi^{(h)}, \ell_\lambda^{(h)} \rangle \right) \quad (7)$$
$$\text{s.t. } \forall h,\ \pi^{(h)} \in \mathcal{U}(\mu_1^{(h)})$$

with $\ell_\lambda^{(h)}(x,y) = -\lambda^\top f^{(h)}(y) - c^{(h)}(x,y)$. For each class $h \in \{1,\ldots,H\}$, each $\lambda \in \mathbf{R}_+^A$, $\varepsilon > 0$, and $x = (s, w_x) \in \mathcal{S}^{(h)} \times \mathcal{W}^{(h)} = \mathcal{X}^{(h)}$, we define:

$$B_{\lambda,\varepsilon}^{(h)}(x) = \varepsilon \log \int \exp\left(\varepsilon^{-1}\ell_\lambda^{(h)}((s, w_x), (s, w_y))\right) \mu_2^{(h)}(s, w_y) dw_y \quad (8)$$

Proposition 1. *(i) The minimum of the dual problem gives:*

$$\forall \lambda \in \mathbf{R}^A, \varphi(\lambda) = -\sum_{h=1}^H \alpha^{(h)} \langle \mu_1^{(h)}, B_{\lambda,\varepsilon}^{(h)} \rangle. \quad (9a)$$

(ii) The optimal coupling $\pi^{(h)}$ satisfies: $\forall x = (s_x, w_x), y = (s_y, w_y) \in \mathcal{X}^{(h)}$

$$\pi_\lambda^{(h)}(x,y) = \mu_1^{(h)}(x)\mu_2^{(h)}(y)\delta_{s_x}(s_y)\exp\left(L_\lambda^{(h)}(x,y)\right), \quad (9b)$$

where: $L_\lambda^{(h)}(x,y) = \varepsilon^{-1}\left\{\ell_\lambda^{(h)}(x,y) - B_{\lambda,\varepsilon}^{(h)}(x)\right\}$
(iii) There is no duality gap.

Proof. As each term in the sum in (7) only depends of $\pi^{(h)}$, the minimization can be performed term by term:

$$\varphi(\lambda) = \sum_{h=1}^H \alpha^{(h)} \min_{\pi^{(h)}} \left(\varepsilon D_{\mathrm{KL}}(\pi^{(h)} \| \mu_1^{(h)} \otimes \mu_2^{(h)}) - \langle \pi^{(h)}, \ell_\lambda^{(h)} \rangle\right)$$

The minimization subproblems were solved in [11] which gives (i),(ii) and (iii). The proof is based on convex duality between relative entropy and log moment generating functions. □

We now introduce the scaled dual function and study its derivative: $\mathcal{J}(\zeta) := -\varepsilon^{-1}\varphi(\varepsilon\zeta)$.

Proposition 2. *Function $\mathcal{J}^{(h)}$ is convex and continuously differentiable, with:*

$$\nabla \mathcal{J}(\zeta) = \sum_{h=1}^H \alpha^{(h)} \langle \mu_\lambda^{(h)}, f^{(h)} \rangle = \sum_{h=1}^H \alpha^{(h)} \mathbb{E}_\lambda^{(h)}[f^{(h)}] \quad (10a)$$

$$\nabla^2 \mathcal{J}(\zeta) = \sum_{h=1}^H \alpha^{(h)} \mathbb{E}_\lambda^{(h)}\left[f^{(h)}(Y)f^{(h)}(Y)^\top - \mathbb{E}_\lambda^{(h)}[f^{(h)}(Y) \mid X]\mathbb{E}_\lambda^{(h)}[f^{(h)}(Y) \mid X]^\top\right] \quad (10b)$$

where $\mu_\lambda^{(h)}$ is the second marginal of $\pi_\lambda^{(h)}$ and $\mathbb{E}_\lambda^{(h)}$ is the expectancy according to $\mu_\lambda^{(h)}$. It implies that $\mathcal{J}$ is strictly convex.

Proof. The proof is a direct differentiation of the dual expressions in Proposition 1. □

2.4 Algorithm

We design a gradient descent algorithm (Algorithm 1), in which $\mathbb{E}_\lambda^{(h)}[f^{(h)}]$ is computed for each class either directly when the state space associated with this class is finite and not too large (as in Sect. 3.2) or through Monte Carlo Methods when this state space is too large or even infinite (as in Sect. 3.3). $\{\rho_k\}_{k=1}^K$ the step size and K the number of gradient descent iterations are chosen empirically to accelerate convergence.

Algorithm 1. a Gradient Descent algorithm to solve MCOT
1: Initialize λ
2: **for** each $k = 1, \ldots, K$ **do**
3: $\quad G \leftarrow \sum_{h=1}^H \alpha^{(h)} \mathbb{E}_\lambda^{(h)}[f^{(h)}]$
4: $\quad \lambda \leftarrow \lambda - \rho_k G$
5: **end for**

3 Model Predictive Control

3.1 The Model Predictive Control Algorithm

Algorithm 2. Model Predictive Control for Heterogeneous loads
1: Initialize $G^{MPC}, \lambda \in \mathbb{R}^T$
2: **for** each $t = 1, \ldots, T$ **do**
3: $\quad$ **for** each $k = 1, \ldots, K$ **do**
4: $\quad\quad G \leftarrow \sum_{h=1}^H \alpha^{(h)} \mathbb{E}_{\lambda_{t:T}}^{(h)}[f_{t:T}^{(h)}]$
5: $\quad\quad \lambda_{t:T} \leftarrow \lambda_{t:T} - \rho_k G$
6: $\quad$ **end for**
7: $\quad G_t^{MPC} \leftarrow \sum_{h=1}^H \alpha^{(h)} \mathbb{E}_{\lambda_{t:t}}^{test,(h)}[f_{t:t}^{(h)}]$
8: **end for**

In the Model Predictive Control (MPC) setting (Algorithm 2), we aim at satisfying some predefined constraints (e.g. tracking or maximum consumption constraints) using prediction of the future dynamic of the loads. For the Water Heaters, the unknowns are the future water drains and for the Electric Vehicles, it is the number of EVs and their charging needs. These unknowns are progressively discovered by the algorithm at each time t, which updates the density considered in the expectancy $\mathbb{E}_\lambda^{(h)}$. More details on this update of the density for each type of agent are presented in the next sections. We denote $f_{t_1:t_2}^{(h)}$, $\lambda_{t_1:t_2}$ the global constraints, respectively the slice of λ from t_1 to t_2, and T the number of time intervals.

3.2 Electric Vehicles

As in [11], we consider a population of Electric Vehicles with their behavior derived from the Elaad OpenDataset [8]. This dataset is composed of 10.000 random transactions from public charging stations operated by EVnetNL in the Netherlands, in the year 2019. We limit ourselves to a typical workday, discretized in $T = 144$ intervals of size $\delta t = 10$ minutes, and separate the dataset between a training set $\mathcal{D}_{train}$ and a test set $\mathcal{D}_{test}$.

Every vehicle is described by its arriving time t_a, its leaving time t_l (which is also the time limit for which it must be fully charged) and its charging needs Δt_n, the duration it needs to be fully charged. Therefore, the state space of all EVs is written as follows:

$$\mathcal{S}^{\text{EVs}} = \underbrace{\{1,\dots,T\}}_{\text{Arriving time } t_a} \times \underbrace{\{1,\dots,T\}}_{\text{Leaving time } t_l} \times \underbrace{\{1,\dots,T\}}_{\text{Charging need } \Delta t_n} \quad (11)$$

We consider here that all EVs have the same power rating, but with different battery levels, represented by the charging needs Δt_n.

Furthermore, we consider that we can control the charging starting time t_c of every EV. At a given time step $t \in \{1,\dots,T\}$, we can define the charging starting time, anywhere in $\mathcal{W}_t^{\text{EVs}} = \{t,\dots,T\}$ if it also satisfies the constraints: $t_a \leq t_c \leq t_l - \Delta t_n$. Finally, we define $\mathcal{X}_t^{\text{EVs}} = \mathcal{S}^{\text{EVs}} \times \mathcal{W}_t^{\text{EVs}}$ the controlled state space. We can now define distributions over $\mathcal{S}^{\text{EVs}}$ and $\mathcal{X}_t^{\text{EVs}}$:

- We define ν as the fraction of vehicles in each state in the training set.

$$\forall s \in \mathcal{S}^{\text{EVs}}, \quad \nu(s) = \frac{\#\{s' \in \mathcal{D}_{train} : s' = s\}}{\#\mathcal{D}_{train}} \quad (12)$$

We denote as $\{s_t^i\} \subset \mathcal{D}_{test}$ the available vehicles that are present and not charging yet.

- We define as ν_t, the empirical fraction of vehicles left to charge:

$$\forall s = (t_a, t_l, \Delta t_n) \in \mathcal{S}^{\text{EVs}}, \quad \nu_t(s) = \begin{cases} \frac{1}{N_t}\sum_i \delta(s - s_t^i) & \text{if } t_a \leq t \\ \frac{N^{\text{EVs}}}{N_t}\nu(s) & \text{if } t_a > t \end{cases} \quad (13)$$

With $N_t = \int_{\mathcal{S}^{\text{EVs}}} \sum_i \delta(s-s_t^i)ds + N^{\text{EVs}} \int_{\mathcal{S}^{\text{EVs}}} \nu(s)\mathbf{1}_{t_a>t}(s)ds$ the normalization factor, counting the number of vehicles already arrived and not charging plus the number of vehicles that are estimated to arrive.

- The nominal distribution $\mu_{1,t}$ is defined by the "Plug when Arrive" strategy:

$$\forall (s, t_c) = (t_a, t_l, \Delta t_n, t_c) \in \mathcal{X}_t^{\text{EVs}}, \quad \mu_{1,t}(s, t_c) = \nu_t(s)\delta(t_c - t_a) \quad (14)$$

- $\mu_{2,t}$ is defined as the "Plug with a uniform distribution" strategy: $\forall s = (t_a, t_l, \Delta t_n, t_c) \in \mathcal{X}_t^{\text{EVs}}$,

$$\mu_{2,t}(s, t_c) = \begin{cases} \frac{1}{t_l - \Delta t_n - \max(t_a,t)+1}\nu_t(s) & \text{if } \max(t_a,t) \leq t_c \leq t_l - \Delta t_n \\ 0 & \text{otherwise} \end{cases} \quad (15)$$

For the MPC algorithm, we define:

- The local cost function c as a quadratic penalization:

$$\begin{aligned} c: \mathcal{X}_t^{\text{EVs}} \times \mathcal{S}^{\text{EVs}} \times \mathcal{W}_t^{\text{EVs}} &\to \mathbb{R} \\ x, (t_a, t_l, \Delta t_n), t_c &\mapsto (t_c - t_a)^2 \end{aligned} \quad (16)$$

c is a penalty for starting charging long after the vehicle arrives.

- We consider that all EVs have the same power consumption of 6kW and define the power (in kW) of one as: $P^{\text{EVs}} : \mathcal{X}_t^{\text{EVs}} \to \mathbb{R}^T$, such that $\forall t' \in \{t,\ldots,T\}$:

$$P_{t'}^{\text{EVs}} : \mathcal{X}_t^{\text{EVs}} \to \mathbb{R} \qquad t_a, t_l, \Delta t_n, t_c \mapsto \begin{vmatrix} 6 \text{ if } t_c \leq t' < t_c + \Delta t_n \\ 0 \text{ otherwise} \end{vmatrix} \qquad (17)$$

3.3 Water Heaters

As in [12], we consider a large population of Water Heaters with their water drains drawn in a dataset from the SMACH (Multi-agent Simulation of Human Activity in the Household) platform [9]. We limit the dataset to working days and separate the dataset between a training set $\mathcal{D}_{train}$ and a test set $\mathcal{D}_{test}$.

At time $t \in [0, 24]$, the i-th WH is modeled by its mean temperature $\theta^i(t) \in \Theta$ and its power mode $m^i(t) \in \{0, 1\}$ (Off/On). These WHs must then follow the *Ordinary Differential Equation* (ODE):

$$\frac{d\theta(t)}{dt} = \underbrace{-\rho(\theta(t) - \theta_{amb})}_{\text{heat loss}} + \underbrace{\sigma m(t)p}_{\text{Joule effect}} - \underbrace{\sigma\epsilon(t)}_{\text{water drain}} \qquad (18)$$

with ρ the fraction of heat loss by minute, σ the specific heat capacity of the volume of water, p the heating power, θ_{amb} the room temperature, and $\epsilon(t)$ the power equivalent of the water drains at time t.

We are interested in controlling their behavior over a day, discretized in $T = 144$ intervals of size $\delta t = 10$ minutes. The previous ODE (18) is thus discretized with the following update:

$$\theta_{t+1} = \theta_t - (\rho(\theta_t - \theta_{amb}) + \sigma m_t p_{max} - \sigma \epsilon_t)\delta t \qquad (19)$$

where for $t \in \{0,\ldots,T\}$, θ_t, m_t and ϵ_t are the values of θ, m and ϵ at time $t \times \delta t$.

We define the nominal dynamics as the uncontrolled dynamic [13,14]. A water heater aims at keeping its mean temperature between $\theta_{min} = 50°C$ and $\theta_{max} = 65°C$ by turning the water heater Off whenever the temperature reaches θ_{max} and turning it back On whenever the temperature reaches below θ_{min}. A nominal trajectory with the drains drawn in $\mathcal{D}_{train}$ is represented in Fig. 1(a).

The full update equation for the nominal policy μ_1 can then be written as follows:

$$\begin{cases} \theta_{t+1} = \theta_t - \rho\delta t(\theta_t - \theta_{amb}) + \sigma\delta t m_t p_{max} - \sigma\epsilon_t \\ m_{t+1} = \begin{cases} m_t & \text{if } \theta_{t+1} \in [\theta_{min}, \theta_{max}] \\ 0 & \text{if } \theta_{t+1} \geq \theta_{max} \\ 1 & \text{if } \theta_{t+1} \leq \theta_{min} \end{cases} \\ \theta_0, m_0 \sim \nu_0 \end{cases} \qquad (20)$$

with ν_0 being the initial density: the initial temperature θ_0 is uniformly distributed in $[\theta_{min}, \theta_{max}]$ and the initial mode m_0 is 1 (On) with probability 0.25 and 0 (Off) with probability 0.75.

Therefore, we can define the state space of all WHs as follows:

$$\mathcal{S}^{\text{WHs}} = \underbrace{\Theta}_{\substack{\text{Initial Temperature} \\ \theta_0}} \times \underbrace{\{0,1\}}_{\substack{\text{Initial mode} \\ m_0}} \times \underbrace{\mathbb{R}_+^T}_{\substack{\text{Random Drains} \\ \{\epsilon_t\}}} \quad (21)$$

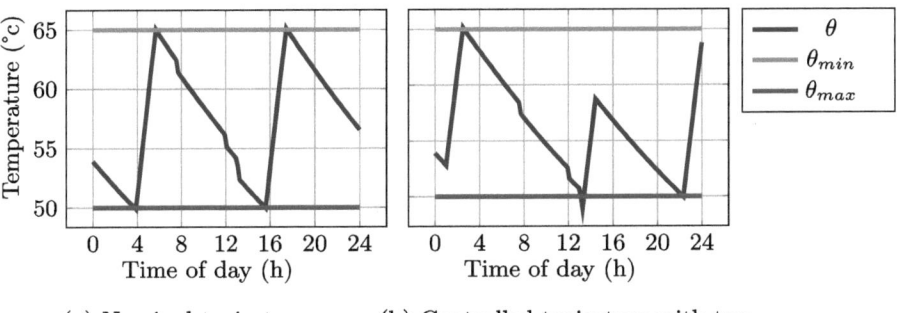

(a) Nominal trajectory (b) Controlled trajectory with two switches: at $t_1 = 1{:}00$ and at $t_2 = 14{:}20$

Fig. 1. Example of trajectory for a single water heater, with $\theta_0 = 54°\text{c}$ and $m_0 = 0$.

Furthermore, we allow the WHs to switch their power mode (from On to Off or from Off to On), while the temperature is still between the two bounds $\theta_{t+1} \in [\theta_{min}, \theta_{max}]$. We limit ourselves to two additional switches per day per WH, compared to its nominal dynamics. We will note these two times $t_1, t_2 \in \{1, \ldots, T\}^2$. This limitation avoids frequent switching, which may be undesirable. Figure 1(b) shows an example of a trajectory with two switches.

Therefore, the power mode update for the policy μ_2 can be written as follows:

$$m_{t+1} = \begin{cases} m_t & \text{if } \theta_{t+1} \in [\theta_{min}, \theta_{max}] \text{ and } t \notin \{t_1, t_2\} \\ (1-m_t) & \text{if } \theta_{t+1} \in [\theta_{min}, \theta_{max}] \text{ and } t \in \{t_1, t_2\} \\ 0 & \text{if } \theta_{t+1} \geq \theta_{max} \\ 1 & \text{if } \theta_{t+1} \leq \theta_{min} \end{cases} \quad (22)$$

We then define $\mathcal{W}_t^{\text{WHs}} = \{t, \ldots, T\}^2 \cup \{t, \ldots, T\} \cup \{\emptyset\}$ the space of at most 2 controlled switches added to a nominal trajectory.

As in Sect. 3.2, we define $\mathcal{X}_t^{\text{WHs}} = \mathcal{S}^{\text{WHs}} \times \mathcal{W}_t^{\text{WHs}}$ the controlled state space. From now on, we consider that $\forall x \in \mathcal{X}_t^{\text{WHs}}$, we can retrieve the two corresponding sequences $\{m_t(x)\}$ and $\{\theta_t(x)\}$ using (20) and (22).

We define the local cost c as the number of added switches:

$$\begin{aligned} c : \mathcal{X}^{\text{WHs}} \times \mathcal{S}^{\text{WHs}} \times \mathcal{W}_t^{\text{WHs}} &\to \mathbb{R} \\ x, s, \text{switches} &\mapsto \#\text{switches} \end{aligned} \quad (23)$$

We consider that all WHs have the same power consumption of 2.2kW and define the power (in kW) of one as: $P^{\text{WHs}} : \mathcal{X}_t^{\text{WHs}} \to \mathbb{R}^T$, such that $\forall t' \in \{t, \ldots, T\}$:

$$\begin{aligned} P_{t'}^{\text{WHs}} : \mathcal{X}_t^{\text{WHs}} &\to \mathbb{R} \\ x &\mapsto 2.2 \times m_{t'}(x) \end{aligned} \quad (24)$$

3.4 Results

We consider a population of $N^{\text{WHs}} = 1000$ WHs with their drains sampled in $\mathcal{D}_{test}^{\text{WHs}}$ and a population of $N^{\text{EVs}} = 1000$ EVs taken in $\mathcal{D}_{test}^{\text{EVs}}$. The proportions are then $\alpha^{\text{EVs}} = \alpha^{\text{WHs}} = 0.5$.

Let $t \in \{1, \ldots, T\}$, a step of the MPC algorithm, let $\{x^i\} \in (\mathcal{X}_t^{\text{WHs}})^{N^{\text{WHs}}}$ and $\{y^j\} \in (\mathcal{X}_t^{\text{EVs}})^{N^{\text{EVs}}}$, the controlled state of both WHs and EVs.

We can define the total power consumption P^{Tot} as follows:

$$\forall t' \geq t, \quad P_{t'}^{\text{Tot}} = \sum_{i=1}^{N^{\text{WHs}}} P_{t'}^{\text{WHs}}(x^i) + \sum_{j=1}^{N^{\text{EVs}}} P_{t'}^{\text{EVs}}(y^j) \quad (25)$$

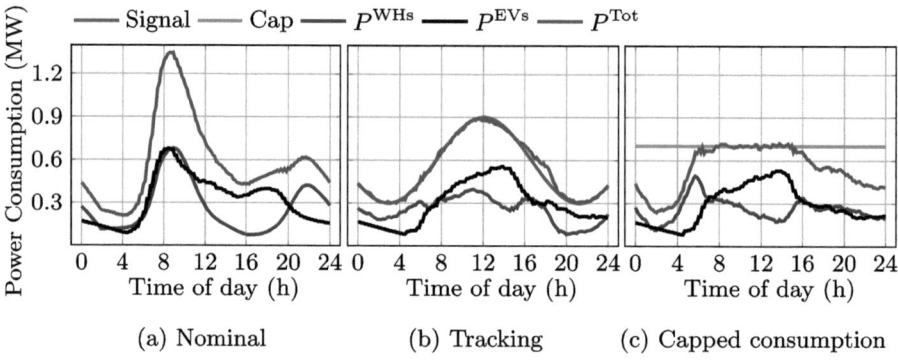

Fig. 2. MPC consumptions compared to the nominal behavior, with tracking constraints in (b) and with consumption cap constraints in (c).

The nominal consumption is obtained using the nominal policy for the EVs and the WHs on a typical workday. We use the above Eq. (25) to get the power consumption during the day, as shown in Fig. 2(a).

Two different types of constraints are considered:

- The tracking constraint, using $\{r_{t'}\}_{t \leq t' \leq T} \in \mathbb{R}^T$ the consumption signal to track in kW. We impose $\forall t' \in \{t, \ldots, T\}, P_{t'}^{\text{Tot}} = r_{t'}$. Using it, Fig. 2(b) shows that we are able to move the consumption peak and reduce it.

- The consumption cap constraint, using $\{u_{t'}\}_{t \leq t' \leq T} \in \mathbb{R}^T$ an upper bound on the consumption. We impose $\forall t' \in \{t, \ldots, T\}, P_{t'}^{\text{Tot}} \leq r_{t'}$. The result is shown in Fig. 2(c), with a total consumption limited to 0.7MW.

To get better results in MPC, we allow ourselves up to 8 switches per WH but still achieve the results with an average of 2 switches per WH during the day.

4 Conclusion and Perspectives

In this work, we have proposed a novel extension of the Moment Constrained Optimal Transport for Control (MCOT-C) framework to accommodate heterogeneous populations of controllable electrical loads. Our formulation allows for the coordination of multiple agent classes, each characterized by distinct dynamics and cost structures, while satisfying global operational constraints such as consumption caps or tracking requirements. By leveraging a mean-field approach with class-wise distributions and introducing an entropic regularization, we derive a tractable dual formulation that admits efficient numerical resolution through gradient-based methods. Furthermore, we combine this approach within a Model Predictive Control (MPC) framework, enabling online adaptation to time-varying or partially revealed information. The practical relevance of our method is demonstrated through numerical experiments on realistic datasets for electric vehicles and water heaters, showing effective constraint enforcement and interpretable decentralized policies.

Several directions for future research naturally emerge from this study. First, a theoretical analysis of the robustness of the MPC strategy to modeling errors or stochastic disturbances would strengthen the practical deployment of the approach. Secondly, the extension to dynamic or time-coupled constraints (such as ramping limitations or time-of-use pricing) may be explored to address more realistic operational scenarios. Finally, the current model relies on historical data to forecast future agent behavior but does not progressively update these forecasts using the information revealed throughout the day. Integrating learning-based methods, such as reinforcement learning, could provide a means to estimate unknown agent models or to improve long-term control performance under uncertainty.

References

1. Carmona, R., Delarue, F.: Probabilistic Theory of Mean Field Games with Applications I-II. Springer, Cham (2018)
2. Achdou, Y., Buera, F.J., Lasry, J.-M., Lions, P.-L., Moll, B.: Partial differential equation models in the socio-economic sciences. Philos. Trans. Roy. Soc. A: Math. Phys. Eng. Sci. **372**, 2014 (2028)
3. Achdou, Y., Lordan, I.: Mean field games with heterogeneity. ESAIM: Proc. Surv. **69**, 1–20 (2020)
4. Cammardella, N., Bušić, A., Meyn, S.: Simultaneous allocation and control of distributed energy resources via Kullback-Leibler-Quadratic optimal control. In: Proceedings of the 2020 American Control Conference (ACC), pp. 514–520. IEEE (2020)
5. Palensky, P., Dietrich, D.: Demand side management: demand response, intelligent energy systems, and smart loads. IEEE Trans. Industr. Inf. **7**(3), 381–388 (2011)
6. Callaway, D.S., Hiskens, I.A.: Achieving controllability of electric loads. Proc. IEEE **99**(1), 184–199 (2011)
7. Alfonsi, A., Coyaud, R., Ehrlacher, V., Lombardi, D.: Approximation of optimal transport problems with marginal moments constraints. Math. Comput. Am. Math. Soc. (2020). https://doi.org/10.1090/mcom/3568

8. Elaad OpenDataset. https://platform.elaad.io/analyses/ElaadNL_opendata.php
9. Albouys, J., Sabouret, N., Haradji, Y., Schumann, M., Inard, C.: SMACH: Multi-agent Simulation of Human Activity in the Household. Springer, Cham (2019)
10. Lasry, J.M., Lions, P.L.: Mean field games. Jpn. J. Math. (2007)
11. Le Corre, T., Bušić, A., Meyn, S.: Moment Constrained Optimal Transport for Control Applications. https://arxiv.org/abs/2208.01958
12. Le Corre, T., Cardinal, J., Bušić, A.: Moment Constrained Optimal Transport for Thermostatically Controlled Loads (2025). https://arxiv.org/abs/2508.20059
13. Moreno, B.M., Brégère, M., Gaillard, P., Oudjane, N.: (Online) Convex Optimization for Demand-Side Management: Application to Thermostatically Controlled Loads. HAL (2023)
14. Cammardella, N., Bušić, A., Meyn, S.: Kullback–Leibler-quadratic optimal control. SIAM J. Control. Optim. **61**(5), 3234–3258 (2023). https://doi.org/10.1137/21M1433654

Expected Extremal Reward of a Markov Decision Process

Olivier Tsemogne(✉) and Yezekael Hayel

Avignon University, Avignon, France
olivier.tsemogne@gmail.com, yezekael.hayel@univ-avignon.fr
https://sites.google.com/view/oliviertsemogne/research-experience ,
https://sites.google.com/site/yezekaelhayelsite/

Abstract. This paper addresses decision-making problems where the objective is to maximize the *best* outcome along a trajectory in a Markov Decision Process (MDP), rather than its cumulative reward. Such extremal objectives naturally arise in risk-sensitive applications, including cybersecurity and resilience planning, but fall outside the scope of classical MDP theory due to their non-Markovian nature. We propose a principled transformation that augments the MDP state space with a deterministic memory variable tracking the maximal reward, yielding an equivalent total-reward MDP solvable by dynamic programming. This construction enables the extraction of Markovian policies through a provision function tailored to each initial state. We evaluate our framework on a malware containment problem, showing that the memory-based baseline policy significantly outperforms a greedy myopic strategy across all configurations. Our results demonstrate improved worst-case cost and containment time, highlighting the utility of memory-augmented planning for extremal performance control in stochastic systems.

Keywords: Markov Decision Processes · Extremal Reward Optimization · Memory-Augmented State · Malware Containment · Risk-Sensitive Planning

1 Introduction

Markov Decision Processes (MDPs) provide a classical framework for sequential decision-making under uncertainty, allowing the optimization of additive criteria such as total, discounted, or average rewards [6]. However, these formulations do not accommodate *extremal* objectives, where the goal is to optimize the maximum (or minimum) value observed along a trajectory. Such objectives arise naturally in settings like cybersecurity or system resilience, where peak costs must be constrained or transient performance must be guaranteed.

Several works have addressed such objectives, for instance through variants of Q-learning adapted to the discounted maximum [7]. More recently, extensions to stochastic environments have been proposed: [2,3] consider deterministic kernels, while [11] introduces an augmented state variable to ensure contractive Bellman-like operators and policy-gradient theorems. However, this construction induces a continuous auxiliary dimension, whose discretization is often required in practice, thus leading to a potential explosion of the effective state space.

Other approaches focus on risk-aware objectives, such as CVaR [1,8] or entropic risk measures [5], sometimes interpreted as smooth approximations of extremes [12]. However, these methods do not provide an exact solution to the problem of optimizing the pointwise extremum along a trajectory.

We propose an exact reformulation of this problem as a *total-MDP*, in which the state is augmented with a memory variable that tracks the maximum value observed so far. Unlike approaches that rely on a continuous auxiliary dimension [11], our construction preserves a discrete and finite augmented state space whenever the original MDP is finite. This structural property controls the state-space explosion and enables the direct application of classical dynamic programming algorithms. Unlike non-Markovian memory-based approaches [4], our formulation ensures an explicit Markovian dynamic. It also differs from [10] by operating in a fully observable, single-agent setting, and by avoiding the need for auxiliary dimensions such as those introduced by k-cutoff mechanisms [9].

Our main contributions are as follows:

- an exact transformation of an extremal-MDP into a total-MDP with memory;
- a provision-based mechanism to extract interpretable Markovian policies from the augmented model;
- a malware containment case study demonstrating the superiority of the derived policy over a myopic baseline in both cost and containment time.

The rest of the paper is organized as follows. Section 2 introduces the φ-MDP framework and highlights the structural properties of the total-MDP used to solve the extremal problem. Section 3 formalizes the Extremal Reward Maximization Problem (ERMP) and presents the associated transformation and its guarantees. Section 4 illustrates the method through a cybersecurity case study. Finally, Sect. 5 discusses limitations and future directions.

2 φ-Markov Decision Process

We introduce a generalized framework for infinite-horizon Markov decision processes, called φ-MDPs, in which the goal is to optimize a user-defined aggregation φ of the reward sequence. This section defines the model, details classical optimization criteria, and discusses the conditions under which optimal policies exist. Our exposition starts with the general formulation and then focuses on the Total Reward Maximization Problem, which will serve as the basis for the subsequent developments.

2.1 Definition

An φ–*Markov Decision Process* (φ-MDP) is defined as any tuple $(\mathcal{S}, \mathcal{A}, P, r, \varphi)$ where:

- $\mathcal{S}$ and $\mathcal{A}$ are finite sets of states and available actions respectively;
- $P \colon \mathcal{S} \times \mathcal{A} \to \Delta(\mathcal{S})$ is the transition function;
- $r \colon \mathcal{S} \times \mathcal{A} \to \mathbb{R}$ is the reward function;
- $\varphi \colon \mathbb{R}^\mathbb{N} \to \mathbb{R}$ is the reward aggregation criterion.

A φ-MDP models the following sequential decision-making problem. At each decision epoch $t \geqslant 1$, a decision-maker selects an action $A^t \in \mathcal{A}$. The system then transitions probabilistically to a new state $S^t \in \mathcal{S}$ and yields a reward $R^t \in \mathbb{R}$. The transition probability is given by: $\mathbb{P}\left(S^t = s' \mid S^{t-1} = s, A^t = a\right) = P(s' \mid s, a)$.

Starting from an initial state S^0, the decision-maker aims to choose actions A^t that maximize the expected aggregate reward $\mathbb{E}[\varphi(\mathbf{R})]$, where $\mathbf{R} = (R^t)_{t \geqslant 1}$ is the sequence of generated rewards. A decision rule at epoch t is a mapping $d^t \colon h^t \mapsto \Delta(\mathcal{A})$, where h^t denotes the history up to epoch t: $h^1 = s^0$, $h^t = \left(s^0, (a^i, s^i)_{i=1}^{t-1}\right)$ for $t \geqslant 2$. A policy is a sequence of decision rules $\pi = (d^t)_{t \geqslant 1}$.

2.2 The Total Reward Maximization Problem

Two classical aggregation criteria are the expected total reward, $\varphi_1(\mathbf{R}) = \sum_{t=1}^{\infty} R^t$, and the expected discounted total reward, $\varphi_\gamma(\mathbf{R}) = \sum_{t=1}^{\infty} \gamma^{t-1} R^t$, for some discount factor $\gamma \in (0,1)$. These criteria are tractable due to the interchangeability of summation and mathematical expectation. We focus on the φ_1-MDP, referred to as the *Total Reward Maximization Problem* (TRMP), or simply the *total-MDP*. In this setting, under appropriate conditions, there exists an optimal policy $\pi^* \in \Pi$, where Π denotes the set of stationary, Markovian, and deterministic policies: $\Pi = \{\pi \colon \mathcal{S} \to \mathcal{A}\}$. Optimality is guaranteed when the MDP satisfies one of the following boundedness conditions, defined in terms of the auxiliary value functions: $V_+^\pi(s) = \mathbb{E}^\pi\left[\sum_{t=1}^{\infty} \max(R^t, 0) \mid S^0 = s\right]$, $V_-^\pi(s) = \mathbb{E}^\pi\left[\sum_{t=1}^{\infty} \max(-R^t, 0) \mid S^0 = s\right]$.

Definition 1. *A model is said to be:*

(a) *Positive bounded if, for all states $s \in \mathcal{S}$:*
  (i) $V_+^\pi(s)$ *is finite for every policy π;*
  (ii) *There exists at least one action $a \in \mathcal{A}$ such that $r(s,a) \geqslant 0$.*
(b) *Negative bounded if:*
  (iii) *For all states $s \in \mathcal{S}$ and all policies π, we have $V_+^\pi(s) = 0$;*
  (iv) *There exists at least one policy π such that $V_-^\pi(s)$ is finite for all $s \in \mathcal{S}$.*

In both cases, standard dynamic programming techniques apply, and the policy iteration algorithm is guaranteed to converge to an optimal policy.

We introduced φ-MDPs and focused on total reward maximization, which admits optimal stationary policies under boundedness conditions. This formulation serves as a foundation for the next section, where we reformulate extremal reward objectives as equivalent total-MDPs.

3 The Extremal Reward Maximization Problem

In this section, we address Markov decision processes where the objective is to optimize an extremal criterion – either the maximum or minimum – of the sequence of step rewards. This setting models situations in which the decision-maker seeks to achieve the best possible outcome (in maximization problems) or to avoid the worst (in minimization problems).

We refer to this problem as the *Extremal Reward Maximization Problem (ERMP)* or simply the *extremal-MDP*, with the criterion given by:

$$\varphi(\mathbf{R}) = \underset{t \geq 1}{\text{ext}}\, R^t,$$

where the operator ext denotes either the maximum or the minimum, depending on the context. The *max-MDP* corresponds to optimizing the best achievable reward, while the *min-MDP*, often via the transformation reward $= -$cost, models the minimization of the worst-case cost.

Since the extremum operator and the mathematical expectation are not generally interchangeable, direct optimization of $\mathbb{E}[\varphi(\mathbf{R})]$ is intractable in general. To address this, we construct an equivalent total-MDP whose solution yields an optimal policy for the original extremal problem.

3.1 From Extremal- to Total-MDP

Given an extremal-MDP $\mathcal{M} = (\mathcal{S}, \mathcal{A}, P, r, \varphi)$, we define the associated total-MDP $\widetilde{\mathcal{M}} = \left(\widetilde{\mathcal{S}}, \mathcal{A}, \widetilde{P}, \widetilde{r}, \widetilde{\varphi}\right)$ as follows:

- $\widetilde{\mathcal{S}} = \mathcal{S} \times M$, where $M = \{r(s,a) \mid (s,a) \in \mathcal{S} \times \mathcal{A}\}$;
- For all $\widetilde{s} = (s, m)$, $\widetilde{s}' = (s', m') \in \widetilde{\mathcal{S}}$ and $a \in \mathcal{A}$:

$$\widetilde{P}(\widetilde{s}' \mid \widetilde{s}, a) = \begin{cases} P(s' \mid s, a) & \text{if } m' = m + \text{ext}\,(r(s,a) - m, 0), \\ 0 & \text{otherwise;} \end{cases} \quad (1)$$

- For all $\widetilde{s} = (s, m) \in \widetilde{\mathcal{S}}$ and $a \in \mathcal{A}$:

$$\widetilde{r}(\widetilde{s}, a) = \text{ext}\,(r(s,a) - m, 0)\,; \quad (2)$$

- $\widetilde{\varphi}$ is defined, for any sequence $(x_t)_{t \geq 1}$, as: $\widetilde{\varphi}\left((x_t)_{t \geq 1}\right) = \sum_{t=1}^{\infty} x_t$, whenever the sum converges.

Each state $\tilde{s} = (s, m)$ encodes both the original state and the current value of the extremum stored in memory. Upon taking action a in state (s, m), the system transitions as in $\mathcal{M}$, but the extremum is updated to reflect any improvement. The immediate reward is the magnitude of this improvement—positive for maximization problems, or negative for minimization problems.

The following proposition shows that $\widetilde{\mathcal{M}}$ is a well-defined total-MDP:

Proposition 1. *1. $\tilde{\mathcal{S}}$ is finite;*
2. For all $(\tilde{s}, a) \in \tilde{\mathcal{S}} \times \mathcal{A}$, $\tilde{P}(\cdot \mid \tilde{s}, a)$ is a probability distribution over $\tilde{\mathcal{S}}$.

Proof. From M is the image of the function r defined on the finite set $\mathcal{S} \times \mathcal{A}$, the sets M then $\tilde{\mathcal{S}}$ are finite. The proof of 2. is: (1) the obvious non-negativity of P, and (2) with $\tilde{s} = (s, m)$, we have:

$$\sum_{\tilde{s}' \in \tilde{\mathcal{S}}} \tilde{P}\left(\tilde{s}' \mid \tilde{s}, a\right) = \sum_{s' \in \mathcal{S}} \tilde{P}\left(s', m + \text{ext}\left(r(s,a) - m, 0\right) \mid \tilde{s}, a\right) = \sum_{s' \in \mathcal{S}} P\left(s' \mid s, a\right) = 1.$$
∎

3.2 Relation Between the Solutions of the Extremal and the Total-MDPs

We now establish a correspondence between the histories, policies, and value functions of the extremal-MDP and its associated total-MDP. The objective in the extremal-MDP is to maximize the extremal step reward – either the maximum or the minimum – relative to a reference value R^0, which is assumed to be already observed or guaranteed to be observed under any policy. The reference value may default to $R^0 = \text{ext}_{(s,a) \in \mathcal{S} \times \mathcal{A}} r(s, a)$, but can be set differently depending on the context. We write $f \succ g$ to mean that f is better than g, i.e., greater in a max-MDP or smaller in a min-MDP.

Setting $m_0 = R^0$, we define a bijective mapping ψ from histories in the extremal-MDP to those in the total-MDP. Let $h = (s_0, (a_i, s_i)_{i=1}^{t-1})$ be a history in the extremal-MDP at decision epoch $t \geq 2$. Its image under ψ is:

$$\psi(h) = \tilde{h} = \left(\tilde{s}_0, (a_i, \tilde{s}_i)_{i=1}^{t-1}\right), \tag{3a}$$

where for $n = 0, \ldots, t - 1$:

$$\tilde{s}_n = (s_n, m_n), \tag{3b}$$
$$m_n = \text{ext}_{i=0,\ldots,n} r_i, \tag{3c}$$
$$r_n = r(s_{n-1}, a_n), \quad \text{for } n \geq 1, \tag{3d}$$
$$r_0 = R^0. \tag{3e}$$

For $t = 1$, we define $\psi(s_0) = (s_0, R^0)$.

Proposition 2. *The mapping ψ is a bijection from the histories of the extremal-MDP to the histories of the total-MDP that begin with memory R^0.*

Each policy π in the extremal-MDP induces a unique policy $\psi\pi$ in the total-MDP defined by:
$$[\psi\pi]\left(\tilde{h}\right) = \pi\left(\psi^{-1}\left(\tilde{h}\right)\right). \tag{4}$$

Proof. Injectivity follows from relation (3b). For surjectivity, note that any history (s, R^0) of the total-MDP at decision epoch 1 is the image of history s. Now take any history $\left(\tilde{s}_0, (a_i, \tilde{s}_i)_{i=1}^{t-1}\right)$, $t \geqslant 2$, of the total-MDP such that $m_0 = R_0$, and let us prove by induction that $m_n = \max_{i=0,\ldots,n} r_i$ for all $n = 0, \ldots, t-1$, where the sequence $(r_i)_{i=0}^{t-1}$ is defined by Eqs. (3c) and (3d).

Case $n = 0$ trivially holds by definition. Suppose the claim holds for some $n \in \{0, \ldots, t-2\}$ and distinguish the cases $r_{n+1} \preccurlyeq m_n$ and $r_{n+1} \succ m_n$. In case $r_{n+1} \preccurlyeq m_n$, (1) gives: $\text{ext}_{i=0,\ldots,n+1} r_i = \text{ext}\left(r_{n+1}, \text{ext}_{i=0,\ldots,n} r_i\right) = \text{ext}(r_{n+1}, m_n) = m_n = m_{n+1}$. In case $r_n \succ m_n$, (2) gives: $\text{ext}_{i=0,\ldots,n+1} r_i = r_{n+1} = m_{n+1}$. ∎

Finally the following lemma help establish a correspondance between the value function of the extremal- and total-MDPs.

Lemma 3. *With $\tilde{r}^0 = r^0$, the sequences $\mathbf{r}$ and $\tilde{\mathbf{r}}$ of rewards associated with the histories h and $\tilde{h}$ satisfies for all $t \geqslant 0$:*
$$m_t = \sum_{n=0}^{t} \tilde{r}^n = \text{ext}_{n=0,\ldots,t} r^n. \tag{5}$$

Proof. By induction. Suppose the equalities hold for some $t \geqslant 0$. (1) and (2) implies that the memory and the rewards evolve according to the relations: $m_{t+1} = m_t + \text{ext}\left(r^{t+1} - m_t, 0\right)$ and $\tilde{r}^{t+1} = \text{ext}\left(r^{t+1} - m_t, 0\right)$. So, if $r^{t+1} \preccurlyeq m_t$, then $m_{t+1} = m_t$ and $\tilde{r}^{t+1} = 0$. Which implies $\sum_{n=0}^{t+1} \tilde{r}^n = \sum_{n=0}^{t} \tilde{r}^n + \tilde{r}^{t+1} = m_t + 0 = m_t = m_{t+1}$ and $\text{ext}_{n=0,\ldots,t+1} r^n = \text{ext}\left(r^{t+1}, \text{ext}_{n=0,\ldots,t} r^n\right) = \text{ext}\left(m_t, r^{t+1}\right) = m_t = m_{t+1}$, hence (5).

If, a contrario, $r^{t+1} \succ m_t$, then $m_{t+1} = r_{t+1}$ and $\tilde{r}^{t+1} = r_{t+1} - m_t$. Which implies $\sum_{n=0}^{t+1} \tilde{r}^n = \sum_{n=0}^{t} \tilde{r}^n + \tilde{r}^{t+1} = m_t + r_{t+1} - m_t = r_{t+1} = m_{t+1}$ and $\text{ext}_{n=0,\ldots,t+1} r^n = \text{ext}\left(r^{t+1}, \text{ext}_{n=0,\ldots,t} r^n\right) = \text{ext}\left(m_t, r^{t+1}\right) = r^{t+1} = m_{t+1}$, hence (5). ∎

Theorem 4. *Both policies π and $\psi\pi$ have the same value function.*

Proof. The relation between the payoffs of trajectories of both MDPs is given in lemma 3 and results in:
$$V^{\psi\pi}(s,m) = \mathbb{E}^{\psi\pi}\left[\sum_{t=0}^{\infty} \widetilde{R}^t \;\middle|\; S^1 = s, R^0 = m\right] = \mathbb{E}^{\pi}\left[\underset{t=0,\ldots,\infty}{\text{ext}} R^t \;\middle|\; S^1 = s\right] = V^{\pi}(s).$$
∎

That is, π is optimal if and only if $\psi\pi$ is optimal.

3.3 Algorithm for the Extremal-MDP

We now construct an algorithm that solves the extremal-MDP by reducing it to an equivalent total-MDP. We begin by establishing the existence of optimal policies in the transformed model.

Theorem 5. *The total-MDP associated with an extremal-MDP admits a stationary, deterministic, and Markovian optimal policy $\widetilde{\pi}^*: \widetilde{\mathcal{S}} \to \mathcal{A}$.*

Proof. . **For the max-MDP** The reward in the associated total-MDP is always an increase in the memory. So, $\widetilde{r}(s,m,a) \geq 0$ for all state and action. This implies point (ii) of Definition 1. The non-negativity of $\widetilde{r}$ also allows for all policy $\widetilde{\pi}$ to write $V^{\widetilde{\pi}}_+(s,m) = \mathbb{E}^{\psi\pi}\left[\sum_{t=0}^{\infty} \widetilde{R}^t \;\middle|\; S^1 = s, R^0 = m\right] =$
$\mathbb{E}^{\pi}\left[\underset{t=0,\ldots,\infty}{\max} R^t \;\middle|\; S^1 = s\right] \leq \underset{(s,a)\in\mathcal{S}\times\mathcal{A}}{\max} r(s,a)$: point (i). So, the total-MDP is a positive bounded model.

. **For the min-MDP** The reward in the associated total-MDP is always a decrease of the memory. So, $\widetilde{r}(s,m,a) \leq 0$ for all state and action. This allows for all policy $\widetilde{\pi}$ to write $V^{\widetilde{\pi}}_+(s,m) = 0$ – point (iii) – and $V^{\widetilde{\pi}}_-(s,m) =$
$\mathbb{E}^{\pi}\left[\underset{t=0,\ldots,\infty}{\min} R^t \;\middle|\; S^1 = s\right] \leq \underset{(s,a)\in\mathcal{S}\times\mathcal{A}}{\min} r(s,a)$ – point (iv) –. So, the total-MDP is a negative bounded model.
∎

The provision value R^0 represents a reward known or assumed to have been realized prior to the planning phase. The goal is to improve upon this value over the remainder of the process. Accordingly, the policy need not adapt to arbitrary initial memory values but only to those that are certainly realized. A *provision function* is any mapping $\rho: \mathcal{S} \to \{r(s,a) \mid a \in \mathcal{A}\}$, assigning to each state an initial extremal memory value. Given a policy $\pi: \widetilde{\mathcal{S}} \to \mathcal{A}$, we define the induced policy on the original state space as: $\pi_\rho(s) = \pi(s,\rho(s))$ This policy is Markovian and deterministic. If ρ is appropriately chosen, then π_ρ is optimal for the extremal-MDP. The complete procedure is outlined in Algorithm 1. The algorithm first constructs the total-MDP (lines 1–10), then derives the optimal extremal-MDP policy π^*_ρ using the provision function ρ (lines 11–13).

4 Numerical Illustrations

We demonstrate our method on a realistic malware containment scenario within a network of five interconnected servers. Each server is modeled as a node in a graph $\mathcal{G} = (\mathcal{V}, \mathcal{E})$, where $\mathcal{V}$ denotes the servers and $\mathcal{E}$ the direct communication links between them. At any time, each node is in one of three states: Susceptible (S), clean and connected; Infectious (I), compromised and connected; or Removed (R), disconnected for patching. The administrator can take action on each node $a_v \in \{0, 1\}$. $a_v = 0$ means no intervention. If the node is connected, $a_v = 1$ triggers a manual disconnection and patching; If already disconnected, $a_v = 1$ reconnects it.

Algorithm 1: Resolution of the extremal-MDP

Input: extremal-MDP $(\mathcal{S}, \mathcal{A}, P, r, \varphi)$
Input: provision function ρ
Output: optimal policy π_ρ^* of the extremal-MDP

1   $M \leftarrow \{r(s,a) : (s,a) \in \mathcal{S} \times \mathcal{A}\}$;
2   $\widetilde{\mathcal{S}} \leftarrow \mathcal{S} \times M$;
3   **for** each s, a, m **do**
4      $\widetilde{r}(s, m, a) \leftarrow \text{ext}(r(s,a) - m, 0)$;
5   **for** each s, s', a, m, m' **do**
6   **if** $m' = m + \widetilde{r}(s, m, a)$ **then**
7          $\widetilde{P}(s', m' | s, m, a) \leftarrow P(s' | s, a)$
8   **else**
9          $\widetilde{P}(s', m' | s, m, a) \leftarrow 0$;
10   $\widetilde{\pi}^* \leftarrow$ optimal policy of total-MDP $\left(\widetilde{\mathcal{S}}, \mathcal{A}, \widetilde{P}, \widetilde{r}, \text{sum}\right)$;
11   $\pi^* \leftarrow \psi \widetilde{\pi}^*$;   /* see equations (3) */
12   **for** each s **do**
13      $\pi_\rho^*(s) \leftarrow \pi^*(s, \rho(s))$;

Figure 1 illustrates this process. Initially (1a), nodes 0, 2, and 4 are infectious, node 3 is susceptible, and node 1 is removed. Action 01100 is applied (1b), reconnecting node 1 and disconnecting node 2. Stochastic dynamics then follow (1c). Each infectious node independently attempts to infect one of its susceptible neighbors with probability p_c, and attempts autonomous recovery with probability p_s. A susceptible node targeted by d infectious neighbors is infected with probability $1 - (1 - p_c)^d$.

The administrator seeks to maximize long-term trust and functionality by minimizing: the *downtime cost*, proportional to the number of removed nodes; the *reputational cost*, proportional to the number of infectious nodes. Letting δ and ρ denote the cost weights, the expected cost of taking action **a** in state

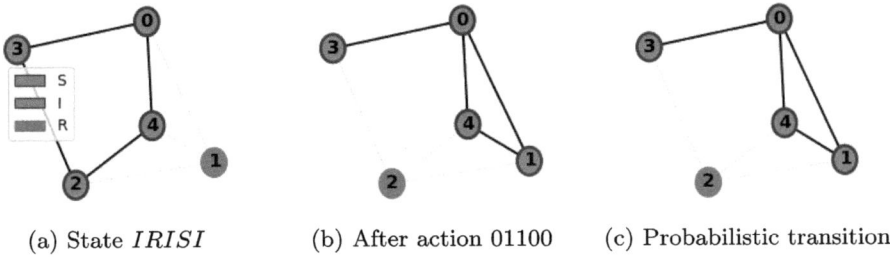

(a) State $IRISI$ (b) After action 01100 (c) Probabilistic transition

Fig. 1. Example of network state evolution with 5 nodes

s is: $C(\mathbf{s}, \mathbf{a}) = \delta \mathbb{E}[\mathcal{I}] + \rho \mathbb{E}[\mathcal{R}]$, where $\mathcal{I}$ and $\mathcal{R}$ denote the number of infectious and removed nodes in the next state. The reward is defined as the negative cost: $R(\mathbf{s}, \mathbf{a}) = -C(\mathbf{s}, \mathbf{a})$.

The resolution of the associated minimal-MDP is performed by transforming it into an equivalent total-MDP, enabling the use of standard solution techniques such as policy iteration. The computed solution specifies, for each network configuration and reward memory, the optimal action to maximize the long-term expected maximum reward.

We instantiate the model using the following parameters: infection success probability $p_c = 0.8$, spontaneous recovery probability $p_s = 0.3$, downtime cost $\delta = 1$, and reputational cost $\rho = 2$. Under these settings, there are 58 different values for the reward, so the transition kernel contains 1,889,568 entries for the minimal-MDP and 6,356,506,752 entries for the associated total-MDP.

To overcome the memory and runtime burden of directly solving the total-MDP with dense matrices, we exploit the sparsity structure of its transition kernel. Rather than storing transition kernel as a large NumPy array, we represent it as a dictionary mapping a single probability value to each 109,594,944 (state, memory, action, state) entries. This data structure significantly reduces overhead and enables dynamic construction of the state-memory-action space during computation. The full resolution of the total-MDP – including the construction of dictionaries for the transition and reward dynamics, and the execution of the policy iteration algorithm – takes 9957.70 s.

We evaluate two Markovian policies. Derived from the extremal-MDP, the *baseline policy* assumes that no cost has yet been incurred: it initializes the provision (i.e., the memory of past rewards) to its lowest possible value, zero, and follows the optimal policy of the total-MDP from there. The *myopic policy*, by contrast, selects at each step the action with the highest immediate reward, ignoring the long-term process. This policy coincides with the greedy choice that minimizes the one-step expected cost.

A comprehensive evaluation was performed. In a *state-wise simulation*, we ran both policies from every one of the 243 network states, over 100 runs of 900 steps each. This setup ensures convergence toward terminal behavior. The result is presented on Fig. 2. In all states where the infection is initially present, the baseline policy consistently outperforms the myopic one – yielding lower maximum costs and faster epidemic resolution. Specifically, the difference in maximum cost is typically between 0 and 1.24 units, with no case where the myopic policy yields better outcomes. The stopping time – the number of steps until the infection disappears – is also significantly shorter under the baseline strategy (with differences reaching up to 65 time steps).

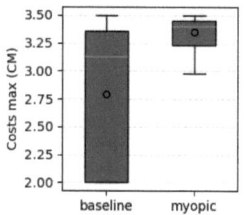 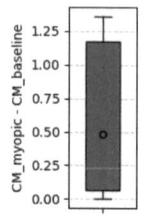 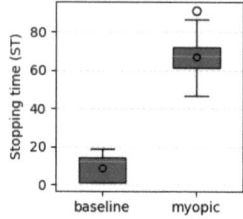

(a) Max cost per policy   (b) Difference in max cost   (c) Stopping time per policy

Fig. 2. State-wise comparison of the *baseline* and *myopic* policies across all initial configurations. *The red dot and the red line respectively represent the mean and the median.* (Color figure online)

In a *fixed-state simulation* from the initial state s_0 = SISSS, 2,000 runs of 1,000 steps each were performed. The *myopic policy* selects the action with the highest immediate reward, leading to an initial cost of 1.5 without disconnecting the infectious node. In contrast, the *baseline policy* immediately performs the disconnection, eradicating the infection in a single step and incurring a maximum cost of 2.0. Under the myopic policy, the infection persists for an average of 61.6 steps, with a maximum cost of 3.22 (90% confidence interval: [3.19, 3.24]). Figure 3 summarizes these differences, showing distributions of cost and stopping time, as well as mean dynamics over time for both strategies.

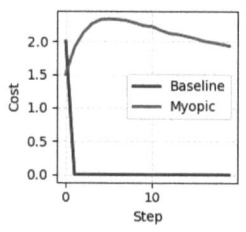 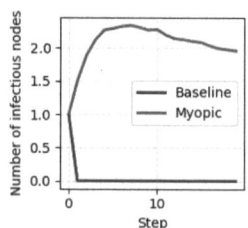 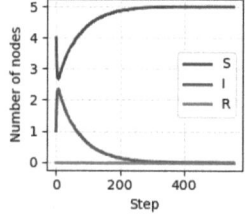

(a) Mean cost over time   (b) Infectious node count   (c) Dynamics under myopic policy

Fig. 3. Detailed simulation from the initial state SISSS. *The baseline policy quickly eliminates the infection and flattens the cost curve, while the myopic policy allows the epidemic to persist longer with higher accumulated cost.*

The baseline policy, even with its simplistic initialization, exploits the structure of the total-MDP to deliver early containment and lower peak impact. In contrast, the myopic strategy, while computationally simpler, fails to anticipate long-term consequences – underscoring the value of temporal foresight in cyber-containment.

5 Conclusion

We proposed a principled framework for solving extremal reward optimization problems in Markov Decision Processes by transforming them into total-reward MDPs with memory-augmented states. This construction enables exact resolution of non-Markovian objectives using classical dynamic programming tools. We illustrated its applicability to malware containment, where our memory-based baseline policy significantly outperforms a myopic alternative in worst-case cost and epidemic stopping time across all network configurations.

Several directions remain open. Our approach assumes known and discrete rewards; extending it to unknown or continuous reward settings would enhance its applicability. Moreover, despite our use of sparse representations, scalability remains a challenge for large-scale networks or long memory horizons. Finally, generalizing the model to more nuanced risk-aware criteria, such as quantile or CVaR objectives, would further strengthen the framework's relevance in safety-critical domains.

Acknowledgment. Research was sponsored by the Army and was accomplished under Grant Number W911NF-24-0149. The views and conclusions contained in this document are those of the authors and should not be interpreted as representing the official policies, either expressed or implied, of the U.S. Army or the U.S. Government. The U.S. Government is authorized to reproduce and distribute reprints for Government purposes notwithstanding any copyright notation herein.

References

1. Chow, Y., Tamar, A., Mannor, S., Pavone, M.: Risk-sensitive and robust decision-making: a cvar optimization approach. In: Advances in Neural Information Processing Systems, vol. 28 (2015)
2. Cui, W., Yu, W.: Reinforcement learning with non-cumulative objective. IEEE Trans. Mach. Learn. Commun. Netw. **1**, 124–137 (2023)
3. Gottipati, S.K., et al.: Maximum reward formulation in reinforcement learning. arXiv preprint arXiv:2010.03744 (2020)
4. Liang, C., Norouzi, M., Berant, J., Le, Q.V., Lao, N.: Memory augmented policy optimization for program synthesis and semantic parsing. In: Advances in Neural Information Processing Systems, vol. 31 (2018)
5. Marthe, A., Bounan, S., Garivier, A., Vernade, C.: Efficient risk-sensitive planning via entropic risk measures. arXiv preprint arXiv:2502.20423 (2025)
6. Puterman, M.L.: Markov Decision Processes: Discrete Stochastic Dynamic Programming. Wiley, New York (2014)

7. Quah, K.H., Quek, C.: Maximum reward reinforcement learning: a non-cumulative reward criterion. Expert Syst. Appl. **31**(2), 351–359 (2006)
8. Rockafellar, R.T., Uryasev, S., et al.: Optimization of conditional value-at-risk. J. Risk **2**, 21–42 (2000)
9. Tomášek, P., Horák, K., Aradhye, A., Bošanskỳ, B., Chatterjee, K.: Solving partially observable stochastic shortest-path games. In: IJCAI-21 Proceedings of the 30th International Joint Conference on Artificial Intelligence, pp. 4182–4189 (2021)
10. Tsemogne, O., Hayel, Y., Kamhoua, C., Deugoué, G.: Optimizing intrusion detection systems placement against network virus spreading using a partially observable stochastic minimum-threat path game. In: Fang, F., Xu, H., Hayel, Y. (eds.) GameSec 2022. LNCS, vol. 13727, pp. 274–296. Springer, Cham (2022). https://doi.org/10.1007/978-3-031-26369-9_14
11. Veviurko, G., Böhmer, W., De Weerdt, M.: To the max: reinventing reward in reinforcement learning. arXiv preprint arXiv:2402.01361 (2024)
12. Zhang, R., Hu, Y., Li, N.: Soft robust MDPS and risk-sensitive MDPS: equivalence, policy gradient, and sample complexity. arXiv preprint arXiv:2306.11626 (2023)

Strategic Interaction Between Queueing System and Impatient User-Base

Anirban Mitra[1], Manu K. Gupta[1(✉)], and N. Hemachandra[2]

[1] Indian Institute of Technology Roorkee, Roorkee-Haridwar Highway, Roorkee 247667, Uttarakhand, India
{anirban_m,manu.gupta}@ms.iitr.ac.in
[2] Indian Institute of Technology Bombay, Powai, Mumbai 400076, Maharashtra, India
nh@iitb.ac.in

Abstract. Many queueing-based service system, such as call centers, telecommunication networks, and ride booking platforms, frequently face user abandonment due to limited service resources. Admission control thus becomes an important strategy for service providers to reject users likely to abandon after entering the system. These policies aim to minimize the long-run average system cost by balancing the costs of rejection, abandonment, and server idling. The quality of service (QoS) experienced by the user-base directly depends on the admission control policy of the service provider. In turn, users respond to QoS experienced by offering an abandonment rate at stationarity. Since the admission control policy also depends on this abandonment rate, a strategic interaction emerges between the service provider and its user-base, which can be modeled as a one-period, two-player non-cooperative game.

This paper investigates the existence of pure-strategy Nash equilibrium (PSNE) in such interactions. Under a mild condition, we prove the monotonicity of PSNE with respect to key system parameters, such as arrival and service rates. In scenarios where a PSNE does not exist, we explore an equilibrium set, which captures cyclical fluctuations of the system between low and high abandonment rate regimes.

Keywords: Queueing systems with abandonment · Admission control · Pure-strategy Nash equilibrium · Equilibrium sets · Quality of service

1 Introduction

Impatient users are an integral aspect of queueing-based service systems[1]. Users are not willing to wait indefinitely for service; they abandon the queue if the

---
[1] Throughout this paper, we use queueing-based service system, service system, and queueing system interchangeably.

waiting time is significantly longer. For instance, a user-base[2] of an online ride booking app may cancel a ride if the estimated wait time is too long. Similarly, in call centers, an impatient caller may leave the virtual queue after prolonged waiting. In telecommunication networks, a subscriber may abandon the queue before a requested connection is fully established. Similarly, an online shopper may abandon a website if the loading time is too long. Many such examples make user abandonment a critical consideration in queueing systems.

Service providers implement admission control policies to selectively restrict arrivals that are likely to abandon the system. These policies minimize the long-run average cost, accounting for costs of user rejection, abandonment, and server idling. Once implemented, users experience a resulting quality of service (QoS) and, in response, offer an abandonment rate at stationarity. Notably, the optimal admission control policy itself depends on this user-set abandonment rate, making the interaction strategic. This interplay can be modeled as a one-period, two-player non-cooperative game between the service provider and its user-base [4] (see Fig. 1).

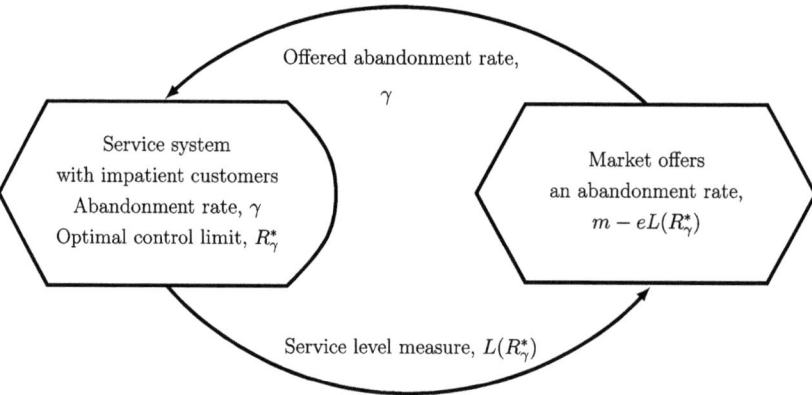

Fig. 1. Strategic interaction between a queueing system and its impatient user-base

In Fig. 1, the queueing system adopts a cost-optimal admission control policy, π^*_γ, in response to a user-specified abandonment rate γ ($\gamma > 0$). This policy results in a certain QoS, denoted by $L(\pi^*_\gamma)$. Based on the QoS experienced, the user-base responds by offering an abandonment rate according to a linear demand model:

$$f(L(\pi^*_\gamma)) := m - eL(\pi^*_\gamma). \tag{1}$$

This affine demand model[3] captures the interaction between the queueing system and the user-base and is adopted here for computational tractability. The

---

[2] We consider an aggregated population of users who are homogeneous in their service requirements, whom we refer as the user-base.

[3] Such models are commonly used in the literature (see [8]) and can be extended to the Cobb-Douglas demand model.

offered abandonment rate lies in the interval $(0, m)$, where m denotes the maximum feasible abandonment rate. In the queue-user interaction function (Eq. (1)), the parameter e represents the elasticity of the abandonment rate with respect to QoS. We examine two key QoS metrics: the fraction of time the system remains idle (L_I), and the fraction of users admitted (L_A). The primary focus of this study is to analyze the equilibrium arising from such strategic interactions, considering the relevant QoS metrics. The main contributions of this paper are summarized below:

- We investigate the existence of the Nash equilibrium for the proposed queue-user interaction, considering the QoS measures L_I and L_A.
- For the QoS measures L_I and L_A, we prove monotonicity of the Nash equilibrium with respect to arrival and service rates under a mild condition.
- We also provide a scheme to compute the Nash equilibrium using best-response dynamics.
- In the absence of a Nash equilibrium, we analyze equilibrium sets and discuss their implications.

1.1 Relevant Literature

Research on admission control in queues with abandonment has received considerable attention in recent years. Hyon and Jean-Marie (2020) [5] examine a single-server system with customer abandonment, where the control decision is whether to serve or idle the server to minimize infinite-horizon discounted cost. A related series of work by Ayhan investigates threshold-based admission control policies. In particular, Ayhan, (2022) [1] examines the optimality of threshold policies in a single-server, single-class queue with customer abandonment. In this model, the system is interested in maximizing the long-run average reward of the system while accounting for holding cost, abandonment cost and service reward. Wu and Ayhan (2025) [10] extend these results to multi-class settings. They obtain the optimal policies in the form of double-set thresholds that converges to single-set threshold under sufficient conditions. Similarly, Koçağa and Ward (2010) [6] study multi-server system with customer abandonment, introducing additional costs such as customer rejection and server idling, and providing diffusion approximations in the Halfin-Whitt regime. Our framework is structurally similar to theirs, but we focus on single-server case to enable a tractable equilibrium analysis, which is substantially more complex in multi-server systems.

Beyond optimal control, several studies address strategic customers in such systems. Benelli and Hassin (2021) [2] analyze threshold-based joining strategies in a queue with abandonment, where customers decide whether to join or balk irreversibly. They show that the individual optimal strategy is independent of the abandonment rate and the socially optimal strategy coincides with the individual one under last-come, first-serve preemptive discipline. Hemachandra ([3,4]) further develops queue-user interaction models and characterizes equilibria arising in such strategic settings. These works are most closely related to ours.

In contrast, our model differs from their study in several key aspects, such as (*i*) considering an *impatient user-base*, (*ii*) considering additional QoS metrics and (*iii*) analyzing the impact of arrival and service rates on Nash equilibrium.

The next section briefly discusses the modeling assumptions.

2 Model Description

This section provides a detailed discussion on the queueing system, some relevant QoS metrics, and the relevant strategic interaction.

2.1 System Description

We consider a single-server queueing system where incoming requests (users) arrive according to a Poisson process with rate λ. A server provides service at an exponentially distributed rate μ. The service provider (controller) decides whether to admit or reject each arriving request based on the system's congestion level (number of users in the system). We assume that each rejecting request incurs a penalty cost c. Once admitted, impatient users may abandon the queue if their waiting time exceeds a certain threshold, resulting in an abandonment cost a. Each user who is not served, abandons the system after an exponentially distributed amount of time with mean $\frac{1}{\gamma}$, where $\gamma > 0$, denotes the abandonment rate. Users who begin service do not abandon. As the server is an expensive resource, idling incurs a cost h_I per unit time, which represents the penalty associated with server underutilization. Figure 2 provides a schematic overview of the system with user abandonment. The service provider seeks to optimize the performance of the system by solving an infinite horizon Markov decision process (MDP) that minimizes the long-term average cost of the system, considering various costs (c, a, and h_I), arrival and abandonment rates.

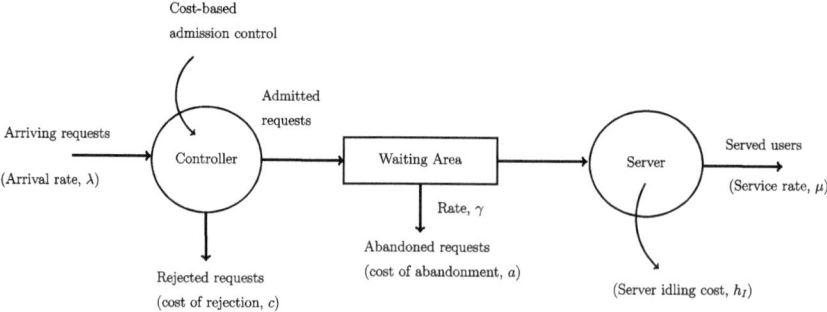

Fig. 2. A schematic view of a queueing system with abandonment.

The optimal control limit R_γ^*, obtained as the solution of the MDP, specifies that the server rejects an arriving user whenever the number of users in the system exceeds this threshold. In our framework, which focuses on threshold-based policies, the optimal policy π^*_γ corresponds to the optimal control limit R^*_γ. We determine the optimal control limit based on the algorithm proposed in Koçağa et al. 2010 [6]. Depending on the parameter space, R^*_γ may remain constant (see Fig. 3, when $\gamma \in [2.74, 10]$) or changes (see Fig. 3, at $\gamma = 2.74$). Average cost functions exhibiting both unique and multiple optimal control limits are presented in Appendix A.1.

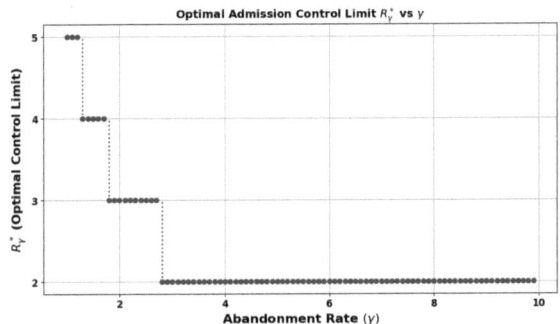

Fig. 3. Parameter space: $\lambda = 5, \mu = 10, c = 1, h_I = 25, a = 5, e = 5, m = 5.6, \gamma \in [2.5, 3.2]$. R^*_γ varies and at some γ there exists maximum two R^*_γ. For example, at $\gamma = 2.74$, $R^*_\gamma = 2$ and 3. Also the long-run average cost is a function of R^*_γ.

2.2 QoS Measures for User-Base

Given an optimal control limit R^*_γ, the user-base experiences a corresponding QoS level. We evaluate two key QoS metrics related to admission-controlled queueing systems with abandonment: the fraction of time the system is idle (L_I), reflecting congestion, and the fraction of users admitted (L_A), indicating service accessibility. Even if some customers subsequently abandon, L_A measures the system's provision of a fair opportunity to be served, which represents a fundamental aspect of service quality. These metrics are derived by modeling the system as a continuous-time Markov chain, with state space representing the number of users in the system. Detailed calculations are provided in Appendix B.

The fraction of time the system is idle,

$$L_I = \left[1 + \frac{1}{\mu} \sum_{j=1}^{R^*_\gamma} \frac{\lambda^j}{\prod_{i=1}^{j-1}(\mu + i\gamma)}\right]^{-1}. \tag{2}$$

The fraction of users admitted,

$$L_A = 1 - \lambda^{R_\gamma^*} \left(\prod_{k=0}^{R_\gamma^*-1} (\mu + k\gamma) \right) \left[1 + \frac{1}{\mu} \sum_{j=1}^{R_\gamma^*} \frac{\lambda^j}{\prod_{i=1}^{j-1}(\mu + i\gamma)} \right]^{-1}. \qquad (3)$$

Next, we model the queue-user strategic interaction as a non-zero sum game.

2.3 The Strategic Model

The interaction between the queueing system and its user-base is strategic (see Fig. 1) and can be modeled as a one-period, two-player non-cooperative game. The players are:

- *Service provider*: chooses a non-negative integer valued admission control limit, $R_\gamma \in \{0, 1, 2, \dots\}$.
- *User-base*: selects an abandonment rate $\gamma \in (0, m)$, according to an interaction function (see Eq. (1)).

A strategy profile (R_γ, γ) is a Nash equilibrium if each player's strategy is a best-response to the other, i.e., $R_\gamma \in b_s(\gamma)$ and $\gamma \in b_u(R_\gamma)$, where b_s and b_u denote the best-response functions of the service provider and the user-base, respectively. A Nash equilibrium occurs when the offered abandonment rate satisfies the fixed-point equation, $\gamma = f(\gamma)$.

Since the service provider's strategy space is discrete, standard equilibrium existence results may not apply. When a pure-strategy Nash equilibrium (PSNE) does not exist, one may consider equilibrium sets or mixed strategies. However, mixed strategies often lack practical relevance, as real-world decisions are shaped by external uncertainties, which render such equilibria less stable and potentially artificial (see [7]). Moreover, Koçağa and Ward (2010) [6] showed (see Theorem 3.1) that no randomized policy can achieve a lower infinite-horizon average cost than a deterministic policy. Therefore, in our model, mixed strategies, such as randomized server actions on whether to accept or reject an arrival, are not well-suited. Next, we discuss PSNE and the impact of some key parameters on it.

3 Pure-Strategy Nash Equilibrium

A PSNE exists if the queue-user interaction converges to a fixed point, i.e., $f(\gamma) = \gamma$. This is ensured when the interaction function $f(\gamma)$ is non-increasing and continuous. Monotonicity guarantees an intersection with the line $f(\gamma) = \gamma$, while continuity prevents the intersection from occurring at a point of discontinuity. Based on this, we formally state two sufficient conditions for the existence of a PSNE.

Condition 1: The interaction function $f(\gamma)$ is non-increasing with respect to the abandonment rate γ.

Condition 2: The interaction function $f(\gamma)$ is continuous over the interval $(0, m)$.

In our model, the interaction function $f(\gamma)$ satisfies both conditions stated above. This is formally established in the following proposition. Proofs of all propositions are provided in Appendix B.

Proposition 1. *Assuming the optimal control limit remains constant over a given parameter space, the interaction function satisfies Conditions 1 and 2 for both QoS measures L_I and L_A, and hence, PSNE exists for this parameter space.*

Justification of the Assumption:

The assumption of a constant optimal control limit is essential, given that a closed-form expression for R_γ^* is generally intractable. This assumption is supported by several numerical examples presented in Appendix A.

Computation of PSNE:

Figure 4 (a) presents a numerical example of the queue-user strategic interaction. Whereas, Fig. 4 (b) illustrates the best-response dynamics corresponding to this interaction. The service provider's objective is to minimize the long-run average cost, formulated as an MDP, as described in Koçağa et al. 2010 [6]. In contrast, the user-base responds through the interaction function $f(\gamma)$ defined in Eq. (1). Starting with an initial abandonment rate γ_0, the service provider solves the MDP to obtain the optimal control limit $R_{\gamma_0}^*$. This, in turn, determines a QoS level $L_A(R_{\gamma_0}^*)$ offered to the user-base. Based on this QoS experienced, the user-base updates its abandonment rate to $f(\gamma_0)$. Next, the service provider solves a new MDP corresponding to $f(\gamma_0)$ and obtains an updated optimal control limit $R_{f(\gamma_0)}^*$, which induces a new QoS level $L_A(R_{f(\gamma_0)}^*)$. Consequently, the user-base responds with a further updated abandonment rate $f(f(\gamma_0))$. This iterative process continues, with both agents responding with their best-strategies, until

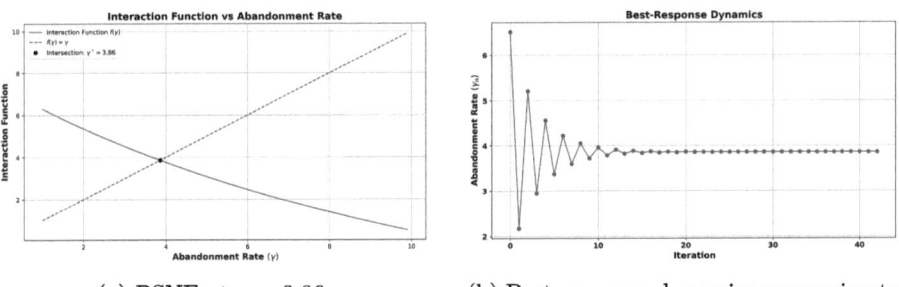

(a) PSNE at $\gamma = 3.86$.

(b) Best-response dynamics converging to $\gamma = 3.86$ (PSNE).

Fig. 4. PSNE and best-response dynamics is illustrated for $\lambda = 15, \mu = 2, c = 1, h_I = 1, a = 5, e = 20, m = 10, \gamma \in [1, 10]$, QoS measure L_A.

convergence is achieved at the PSNE, $\gamma = 3.86$ (rounded down to two decimal places.).

Next, we investigate the impact of arrival rate, service rate, and cost parameters on PSNE.

3.1 Sensitivity Analysis of PSNE

This section studies how arrival rate (λ), service rate (μ), and cost parameters (c, a, h_I) affect PSNE, assuming the optimal control limit (R_γ^*) remains constant over the parameter space. While R_γ^* lacks a closed-form expression, this assumption aids analytical tractability and is supported by numerical examples (see Fig. 3, $\gamma \in [2.74, 10]$). The following proposition states the monotonicity of PSNE with respect to λ and μ.

Proposition 2. *For both the QoS measures L_I and L_A, PSNE increases monotonically with respect to arrival rate and decreases monotonically with respect to service rate.*

Implications of Proposition 2:

Considering a constant R_γ^*, an increase in arrival rate (λ) raises congestion, reducing system idleness (L_I) and prompting users to adopt a higher abandonment rate (γ). Conversely, a higher service rate (μ) increases idleness, resulting in users to lower γ.

Similarly, from the service provider's perspective, higher λ leads to more frequent rejections, lowering L_A and further encouraging higher user abandonment. In contrast, higher μ improves L_A, eliciting a lower γ from users. Proposition 2 is numerically validated in Appendix A.

Now, the impact of cost parameters on PSNE is presented in the following remark.

Remark 1. If the optimal control limit R_γ^* remains constant over a given parameter space, then the cost of rejection (c), the cost of abandonment (a), and the cost of server idling (h_I) have no impact on the PSNE.

This is because the QoS metrics L_I and L_A are independent of c, a, and h_I. These cost parameters only affect the determination of R_γ^*, which is assumed to be fixed in this context. The next section addresses complex cases where a PSNE does not exist.

4 Equilibrium Sets

In the previous section, we examined PSNE of the queue-user strategic interaction. While Condition 2 assumes the interaction function $f(\gamma)$ is continuous over $(0, m)$, certain parameter settings can lead to a piecewise discontinuity in $f(\gamma)$. For instance, in Fig. 3, when $\gamma = 2.74$, the optimal control limit R_γ^* shifts abruptly from 3 to 2. Since $f(\gamma)$ is defined in terms of R_γ^* (see Eq. (1)), this abrupt change results in a discontinuity in $f(\gamma)$. If the line $f(\gamma) = \gamma$ intersects a discontinuity, no fixed point exists. In such cases, we explore the concept of the equilibrium set. We define an *equilibrium set* E and *repelling set* R as in [3] as follows:

If there exist some $\gamma_0 \in (0, m)$ and $E := E_1 \cup E_2$ where $E_1 := \left(0, f(\gamma_0)\right]$, $E_2 := \left[\lim_{\gamma \to \gamma_0^+} f(\gamma), m\right)$, and $R := \left(f(\gamma_0), \lim_{\gamma \to \gamma_0^+} f(\gamma)\right)$ such that

- $f(\gamma) \in E$ for $\gamma \in R$, and,
- $f(\gamma) \in E$ for $\gamma \in E$ ($f(\gamma) \in E_2$ for $\gamma \in E_1$ and $f(\gamma) \in E_1$ for $\gamma \in E_2$).

An equilibrium set E is defined such that the user-base's best response remains within the set, i.e., $f(\gamma) \in E$ for all $\gamma \in E$. In contrast, a repelling set R is one where the best response lies outside the set, indicating that strategies in R tend to move toward the equilibrium set E.

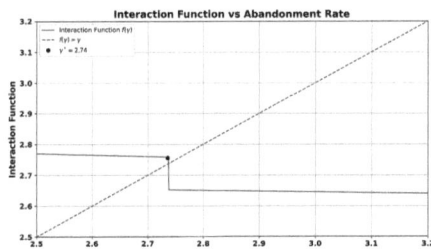
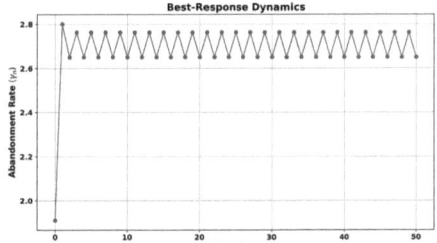

(a) Equilibrium set for the queue-user strategic interaction.

(b) Best-response dynamics oscillates.

Fig. 5. Parameter space: $\lambda = 5, \mu = 10, c = 1, h_I = 25, a = 5, e = 5, m = 5.6$, QoS measure L_I.

In Fig. 5 (a), the $f(\gamma) = \gamma$ line passes through a discontinuity of $f(\gamma)$. Hence, the queue-user strategic interaction is unable to converge to a fixed point. In this example, $\gamma_0 = 2.74, f(\gamma_0) = 2.65, \lim_{\gamma \to \gamma_0^+} f(\gamma) = 2.75$. Hence, $E_1 = (0, 2.65]$, $E_2 = [2.75, 5.6)$, equilibrium set, $E = E_1 \cup E_2 = (0, 2.65] \cup [2.75, 5.6)$; repelling set, $R = (2.65, 2.75)$. The values comprising E and R are rounded down to two decimal places. For this strategic interaction, the queueing system fluctuates between low and high abandonment rate regimes (see Fig. 5 (b)).

Suppose the user-base initially selects a high abandonment rate $\gamma_0 \in E_2$, leading to more frequent user abandonment and inefficient resource use. This degrades the QoS, prompting the user-base to reduce its abandonment rate to $\gamma_1 := f(\gamma_0) \in E_1$. It leads to the improvement in QoS and then incentivizes the user-base to increase the abandonment rate, $\gamma_2 := f(\gamma_1) \in E_2$, and the cycle repeats with $\gamma_3 := f(\gamma_2) \in E_1$, and so on. This alternating pattern reflects persistent oscillations of the queueing system between high and low abandonment rate regimes. The equilibrium set characterizes such cyclical dynamics arising from mutual strategic responses of the service provider and the user-base. In Fig. 5(b), the abandonment rate γ evolves as follows: at iteration 0, $\gamma = 1.90 \in E_1$; iteration 1, $\gamma = 2.80 \in E_2$; iteration 2, $\gamma = 2.65 \in E_1$; iteration 3, $\gamma = 2.76 \in E_2$; iteration 4, $\gamma = 2.65 \in E_1$; iteration 5, $\gamma = 2.76 \in E_2$; and so forth. This sequence demonstrates that the system oscillates between the low abandonment regime E_1 and the high abandonment regime E_2.

5 Summary and Future Research Directions

This paper examines the strategic interaction between a cost-minimizing queueing system with abandonment and a QoS-sensitive user-base. This study identifies PSNE under various QoS measures. We analyze the sensitivity of PSNE with respect to arrival rate, service rate, and cost parameters. We also analyze equilibrium sets when PSNE does not exist.

Future work may extend this framework to general service time distributions with multi-server settings. One may consider several other QoS measures, e.g., the fraction of users abandoned, and analyze the equilibrium behavior of the system. Investigating sufficient conditions for equilibrium set existence, the monotonicity of PSNE under changing optimal control limits and the key cost parameters, present promising directions. Further research could also explore discounted-cost models that offer richer insights into strategic dynamics in real-world service systems.

A Appendix: Numerical Examples

A.1 Cost Functions For Unique and Multiple R^*_γ

For a given parameter space, the long-run average cost is minimized for the optimal control limit R^*_γ. Figure 6 illustrates a unique R^*_γ and Fig. 7 illustrates multiple R^*_γ. A discontinuity in R^*_γ at a specific γ indicates the existence of multiple optimal control limits, the previous and the current one (see Fig. 3, at $\gamma = 2.74$).

A.2 Some Numerical Examples of Constant R^*_γ

This section presents numerical examples illustrating cases where the optimal control limit R^*_γ remains constant over a given parameter space (Fig. 8).

Table 1 illustrates some more examples of constant R^*_γ.

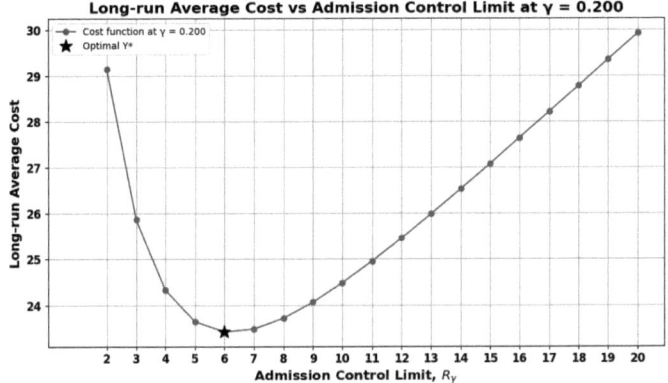

Fig. 6. Parameter space: $\lambda = 60, \mu = 40, \gamma = 0.2, c = 1, h_I = 2, a = 4$. The long-run average cost is minimized at $R_\gamma^* = 6$.

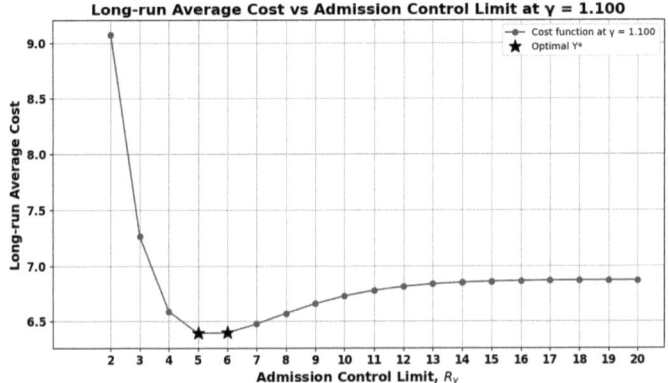

Fig. 7. Parameter space: $\lambda = 30, \mu = 40, \gamma = 1.1, c = 1, h_I = 2, a = 4$. The long-run average cost is minimized at $R_\gamma^* = 5, 6$.

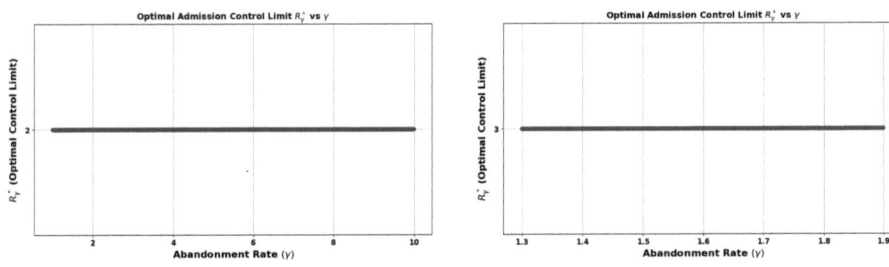

(a) Parameter Space: $\lambda = 50, \mu = 4, c = 1, h_I = 2, a = 3, \gamma \in [1, 10]$.

(b) Parameter Space: $\lambda = 2, \mu = 6, c = 1, h_I = 2, a = 2.5, \gamma \in [1.3, 1.9]$.

Fig. 8. (a) Parameter Space: $\lambda = 50, \mu = 4, c = 1, h_I = 2, a = 3, \gamma \in [1, 10]$. (b) Parameter Space: $\lambda = 2, \mu = 6, c = 1, h_I = 2, a = 2.5, \gamma \in [1.3, 1.9]$.

Table 1. Parameter Spaces For Fixed R_γ^*

Experiment	λ	μ	c	a	h_I	γ	Constant R_γ^*
1	200	300	10	25	5	[15, 17]	6
2	500	450	100	250	50	[15, 19]	5
3	1	2	5	8	2	[0.6, 0.8]	4
4	10	20	50	100	25	[0.78, 0.84]	14
5	550	420	10	22	5.5	[7.5, 10]	6
6	330	150	12	25	5	[5, 10]	14

Table 2. Impact of λ on PSNE. The fixed parameters: $\mu = 3, c = 10, h_I = 50, a = 200, e = 15, m = 12, \gamma \in [0.5, 10]$, QoS is L_I.

Arrival Rate	PSNE
1	1.36007
1.2	1.97632
1.4	2.53456
1.6	3.04241
1.8	3.50621
2	3.93128
2.2	4.32214
2.4	4.68262
2.6	5.01601
2.8	5.32515

Table 3. Impact of μ on PSNE. The fixed parameters: $\lambda = 5, c = 8, h_I = 70, a = 1500, e = 25, m = 15, \gamma \in [1, 10]$, QoS is L_I.

Service Rate	PSNE
2	9.54475
2.2	9.15052
2.4	8.77419
2.6	8.41425
2.8	8.06934
3	7.73827
3.2	7.41997
3.4	7.11349
3.6	6.81796
3.8	6.53262

A.3 Numerical Illustration of Proposition 2

In this section, we numerically illustrate Proposition 2.

In Tables 2 and 3, we consider L_I as a QoS measure. Table 2 illustrates a monotonic increasing nature of PSNE with respect to λ. In contrast, Table 3, illustrates a monotonic decreasing nature of PSNE with respect to μ.

In Tables 4 and 5, we consider L_A as a QoS measure. The results are similar to Tables 2 and 3.

B Appendix: Proofs of QoS Measures and Propositions

B.1 Computing the QoS Measures L_I and L_A

We aim to evaluate two key performance measures in a single-server queueing system with abandonment: the fraction of time the system is idle (L_I) and the fraction of users admitted (L_A). To achieve this, we model the system as a continuous-time Markov chain (CTMC), where:

Table 4. Impact of λ on PSNE. The fixed parameters: $\mu = 20, c = 10, h_I = 10, a = 50, e = 20, m = 20, \gamma \in [2, 10]$, QoS is L_A.

Arrival Rate	PSNE
10	2.57334
12	3.24395
14	3.88813
16	4.49923
18	5.07499
20	5.61553
22	6.12220
24	6.59694
26	7.04193
28	7.45938

Table 5. Impact of μ on PSNE. The fixed parameters: $\lambda = 50, c = 12, h_I = 40, a = 65, e = 35, m = 20, \gamma \in [1, 15]$, QoS is L_A.

Service Rate	PSNE
5	10.92093
7	10.15807
9	9.40226
11	8.65524
13	7.91855
15	7.19358
17	6.48155
19	5.78353
21	5.10047
23	4.43313

- The states represent the number of users in the system, denoted as n.
- The birth rate (arrival rate) is given by λ.
- The state-dependent death rate accounts for both service completion and user abandonment, formulated as:

$$\mu[n \wedge 1] + \gamma[n-1]^+,$$

where μ is the service rate and γ is the abandonment rate.

Fig. 9 illustrates the structure of this CTMC.

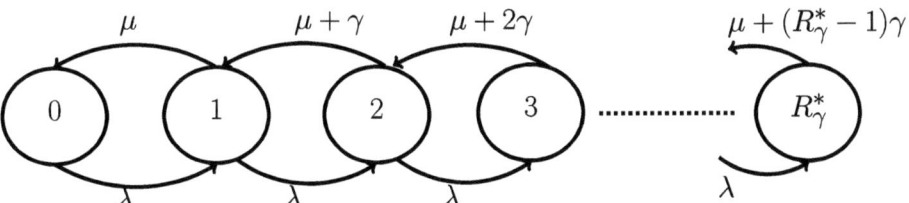

Fig. 9. In this CTMC, state space is the number of users in the system (n), birth rate is λ and the state-dependent death rate is $\mu[n \wedge 1] + \gamma[n-1]^+$

The stationary probability that the system has n number of users when R_γ^* is the optimal control limit is given bellow and that can be computed by solving the given stationary distribution (see Fig. 9).

$$p_n = \frac{\lambda^n}{\prod_{k=0}^{n-1}(\mu+k\gamma)\left[1+\frac{1}{\mu}\sum_{j=1}^{n}\frac{\lambda^j}{\prod_{i=1}^{j-1}(\mu+i\gamma)}\right]}, \quad n=1,\cdots,R_\gamma^*.$$

Also, the stationary probability that the system is empty is

$$p_0 = \frac{1}{\left[1+\frac{1}{\mu}\sum_{j=1}^{R_\gamma^*}\frac{\lambda^j}{\prod_{i=1}^{j-1}(\mu+i\gamma)}\right]}.$$

Hence, the fraction of time the system is idle is,

$$L_I = p_0 = \frac{1}{\left[1+\frac{1}{\mu}\sum_{j=1}^{R_\gamma^*}\frac{\lambda^j}{\prod_{i=1}^{j-1}(\mu+i\gamma)}\right]}.$$

We use the concept of Poisson Arrival See Time Averages (see [9]) to determine the stationary probability that arriving users see R_γ^* number of users in the system. Hence, the required stationary probability is

$$p_{(n=R_\gamma^*)} = \frac{\lambda^{R_\gamma^*}}{\prod_{k=0}^{R_\gamma^*-1}(\mu+k\gamma)\left[1+\frac{1}{\mu}\sum_{j=1}^{R_\gamma^*}\frac{\lambda^j}{\prod_{i=1}^{j-1}(\mu+i\gamma)}\right]}.$$

Hence, the fraction of users admitted is

$$L_A = 1 - \frac{\lambda^{R_\gamma^*}}{\prod_{k=0}^{R_\gamma^*-1}(\mu+k\gamma)\left[1+\frac{1}{\mu}\sum_{j=1}^{R_\gamma^*}\frac{\lambda^j}{\prod_{i=1}^{j-1}(\mu+i\gamma)}\right]}.$$

B.2 Proof of Proposition 1

We consider that the optimal control limit remains constant over a given parameter space.

Let's validate whether $f(\gamma)$ satisfies Condition 1 for L_I and L_A. We first consider L_I as the QoS measure. From Eq. (1), we can express:

$$\frac{\partial f(\gamma)}{\partial \gamma} = -e \frac{\partial L_I}{\partial \gamma}. \tag{4}$$

From Eq. (2), we can write:

$$\frac{\partial L_I}{\partial \gamma} = -\frac{1}{u^2} \frac{\partial u}{\partial \gamma}, \tag{5}$$

where,

$$u = 1 + \frac{\lambda}{\mu} + \frac{\lambda^2}{\mu}(\mu+\gamma)^{-1} + \frac{\lambda^3}{\mu}[(\mu+\gamma)(\mu+2\gamma)]^{-1} + $$
$$\cdots + \frac{\lambda^{R_\gamma^*}}{\mu}[(\mu+\gamma)(\mu+2\gamma)\cdots\{\mu+(R_\gamma^*-1)\gamma\}]^{-1}.$$

This expression can be rewritten as follows:

$$u = 1 + \frac{\lambda}{\mu} + \frac{\lambda^2}{\mu}f_1(\gamma) + \frac{\lambda^3}{\mu}f_1(\gamma)f_2(\gamma) + \cdots + \frac{\lambda^{R_\gamma^*}}{\mu}f_1(\gamma)f_2(\gamma)\cdots f_{R_\gamma^*-1}(\gamma),$$

where,

$$f_1(\gamma) = (\mu+\gamma)^{-1}, \quad f_2(\gamma) = (\mu+2\gamma)^{-1}, \quad f_3(\gamma) = (\mu+3\gamma)^{-1},$$
$$\cdots, \quad f_{R_\gamma^*-1} = \{\mu+(R_\gamma^*-1)\gamma\}^{-1}.$$

Note that

$$f_1(\gamma), f_2(\gamma), \cdots, f_{R_\gamma^*-1}(\gamma) > 0,$$

and,

$$\frac{\partial f_1(\gamma)}{\partial \gamma}, \frac{\partial f_2(\gamma)}{\partial \gamma}, \frac{\partial f_3(\gamma)}{\partial \gamma}, \cdots, \frac{\partial f_{R_\gamma^*-1}(\gamma)}{\partial \gamma} > 0.$$

Hence, $\prod_{j=1}^{n} \frac{\partial f_j(\gamma)}{\partial \gamma} > 0$, $\forall\, n \in \{2, R_\gamma^* - 1\}$. Therefore,

$$\frac{\partial u}{\partial \gamma} = \frac{-\lambda^2}{\mu(\mu+\gamma)^2} + \frac{-\lambda^3[2(\mu+\gamma)+(\mu+2\gamma)]}{[(\mu+\gamma)(\mu+2\gamma)]^2} +$$

$$\cdots + \frac{\lambda^{R^*_\gamma} \frac{\partial}{\partial\gamma}[(\mu+\gamma)(\mu+2\gamma)\cdots\{\mu+(R^*_\gamma-1)\}]}{\mu[(\mu+\gamma)(\mu+2\gamma)\cdots\{\mu+(R^*_\gamma-1)\}]^2} < 0.$$

From Eq. (4) and Eq. (5), we can write $\dfrac{\partial L_I}{\partial \gamma} > 0$ and $\dfrac{\partial f(\gamma)}{\partial \gamma} < 0$.

Next, we consider the fraction of users admitted, denoted as L_A, as a QoS measure. From Eq. (1), we can express it as follows:

$$\frac{\partial f(\gamma)}{\partial \gamma} = -e \frac{\partial L_A}{\partial \gamma}. \tag{6}$$

From Eq. (3), we can write as follows:

$$\frac{\partial L_A}{\partial \gamma} = \frac{\lambda^{R^*_\gamma}}{v^2} \frac{\partial v}{\partial \gamma}, \tag{7}$$

where,

$$v = [\mu(\mu+\gamma)(\mu+2\gamma)\cdots\{\mu+(R^*_\gamma-1)\gamma\}] + \lambda[(\mu+\gamma)(\mu+2\gamma)(\mu+3\gamma)$$
$$\cdots\{\mu+(R^*_\gamma-1)\gamma\}] + \lambda^2[(\mu+2\gamma)(\mu+3\gamma)(\mu+4\gamma)\cdots\{\mu+(R^*_\gamma-1)\gamma\}] + \cdots + \lambda^{R^*_\gamma}.$$

This expression can be written as follows:

$$v = \mu\psi_1(\gamma)\psi_2(\gamma)\cdots\psi_{R^*_\gamma-1}(\gamma) + \lambda[\psi_1(\gamma)\psi_2(\gamma)\cdots\psi_{R^*_\gamma-1}(\gamma)] +$$
$$\lambda^2[\psi_2(\gamma)\psi_3(\gamma)\cdots\psi_{R^*_\gamma-1}(\gamma)] + \cdots + \lambda^{R^*_\gamma},$$

where,

$\psi_1(\gamma) = (\mu+\gamma)$, $\psi_2(\gamma) = (\mu+2\gamma)$, $\psi_3(\gamma) = (\mu+3\gamma)$, $\cdots$, $\psi_{R^*_\gamma-1}(\gamma) = \{\mu+(R^*_\gamma-1)\gamma\}$.

Note that

$$\psi_1(\gamma), \psi_2(\gamma), \cdots, \psi_{R^*_\gamma-1}(\gamma) > 0,$$

and,

$$\frac{\partial \psi_1(\gamma)}{\partial \gamma}, \frac{\partial \psi_2(\gamma)}{\partial \gamma}, \frac{\partial \psi_3(\gamma)}{\partial \gamma}, \cdots, \frac{\partial \psi_{R^*_\gamma-1}(\gamma)}{\partial \gamma} > 0.$$

Hence, $\prod_{j=1}^{n} \frac{\partial \psi_j(\gamma)}{\partial \gamma} > 0$, $\forall\, n \in \{2, R^*_\gamma - 1\}$. We further observe from Eq. (7) that $\frac{\partial L_A}{\partial \gamma} > 0$ and from Eq. (6), that $\frac{\partial f(\gamma)}{\partial \gamma} < 0$. Hence, the interaction function $f(\gamma)$ satisfies Condition 1 for both the QoS measures L_I and L_A.

Let's validate whether $f(\gamma)$ is continuous on the interval $(0, m)$ for the QoS measures L_I and L_A. We first consider L_I as a QoS measure.

The fraction of time the system is idle,

$$L_I = \frac{1}{\left[1 + \frac{1}{\mu} \sum_{j=1}^{R^*_\gamma} \frac{\lambda^j}{\prod_{i=1}^{j-1}(\mu + i\gamma)}\right]}.$$

Let's say $f_j(\gamma) = \frac{\lambda^j}{\prod_{i=1}^{j-1}(\mu + i\gamma)}$, for $j = 1, 2, \cdots, R^*_\gamma$. Each term of the denominator of $f_j(\gamma)$ is linear function of γ. Hence, all these terms are continuous. Since, $\mu, \gamma, m > 0$, therefore, the product, $\prod_{i=1}^{j-1}(\mu + i\gamma) > 0$ and continuous. Hence, $f_j(\gamma)$ is also continuous. We now define, $f_k(\gamma) = \sum_{j=1}^{R^*_\gamma} f_j(\gamma)$. Since $f_j(\gamma)$ is continuous, their finite sum is also continuous. Let's say $g(\gamma) = 1 + \frac{1}{\mu} f_k(\gamma)$. This is a sum and a scalar product of a continuous function, which is continuous. Note that $f_k(\gamma) > 0$ and hence, $g(\gamma) > 0$. Therefore, $L_I = \frac{1}{g(\gamma)}$ is continuous on the interval $(0, m)$, within which R^*_γ is fixed. We now consider L_A as a QoS measure.

The fraction of users admitted,

$$L_A = 1 - \frac{\lambda^{R^*_\gamma}}{\prod_{k=0}^{R^*_\gamma - 1}(\mu + k\gamma)\left[1 + \frac{1}{\mu}\sum_{j=1}^{R^*_\gamma} \frac{\lambda^j}{\prod_{i=1}^{j-1}(\mu + i\gamma)}\right]}.$$

For the QoS measure L_I We have already shown that $\prod_{k=0}^{R^*_\gamma - 1}(\mu + k\gamma)$

and $\left[1 + \dfrac{1}{\mu}\sum_{j=1}^{R^*_\gamma} \dfrac{\lambda^j}{\prod_{i=1}^{j-1}(\mu+i\gamma)}\right]$ both are continuous functions. Hence, their product $v_j(\gamma)$ is also continuous. Also, $v_k(\gamma) = \dfrac{\lambda^{R^*_\gamma}}{v_j(\gamma)}$ is also continuous, as $v_j(\gamma) > 0$. Hence, $L_A = 1 - v_k(\gamma)$ is continuous on the interval $(0, m)$, within which R^*_γ is fixed.

This completes the proof.

B.3 Proof of Proposition 2

We first examine the impact of λ on PSNE, considering L_I as a QoS measure. Let's say for a given parameter space, PSNE exists at γ^*_1 and for another parameter space, PSNE exists at γ^*_2. Hence, the conditions for the existence of the PSNEs are: $\gamma^*_2 = m - eL_I(\lambda_2)$ and $\gamma^*_1 = m - eL_I(\lambda_1)$. Now, $\gamma^*_2 - \gamma^*_1 = e[L_I(\lambda_1) - L_I(\lambda_2)]$. Let's compare the denominators of $L_I(\lambda_1)$ and $L_I(\lambda_2)$. It turns out to be as follows:

$$\dfrac{1}{\mu}\left[(\lambda_1 - \lambda_2) + \dfrac{1}{(\mu+\gamma)}(\lambda_1^2 - \lambda_2^2) + \dfrac{1}{(\mu+\gamma)(\mu+2\gamma)}(\lambda_1^3 - \lambda_2^3) + \right.$$
$$\left. \cdots + \dfrac{1}{(\mu+\gamma)(\mu+2\gamma)\cdots\{\mu+(R^*_\gamma-1)\gamma\}}(\lambda_1^{R^*_\gamma} - \lambda_2^{R^*_\gamma})\right].$$

Clearly, Denominator of $L_I(\lambda_1)$ − Denominator of $L_I(\lambda_2) < 0$ as $\lambda_2 > \lambda_1$. As L_I is the reciprocal of its denominator, hence, $L_I(\lambda_1) > L_I(\lambda_2)$ and it leads to $\gamma^*_2 > \gamma^*_1$. This completes the first portion of the proof.

Let's examine the impact of μ on PSNE. In this case, the PSNE conditions are: $\gamma^*_2 = m - eL_I(\mu_2)$ and $\gamma^*_1 = m - eL_I(\mu_1)$. Similar to the earlier case, we observe $\gamma^*_2 - \gamma^*_1 = e[L_I(\mu_1) - L_I(\mu_2)]$. Let's compare the denominators of $L_I(\mu_1)$ and $L_I(\mu_2)$. It turns out to be as follows:

$$\lambda\left[\dfrac{1}{\mu_1} - \dfrac{1}{\mu_2}\right] + \lambda^2\left[\dfrac{1}{\mu_1(\mu_1+\gamma)} - \dfrac{1}{\mu_2(\mu_2+\gamma)}\right] + \lambda^3\left[\dfrac{1}{\mu_1(\mu_1+\gamma)(\mu_1+2\gamma)} - \dfrac{1}{\mu_2(\mu_2+\gamma)(\mu_2+2\gamma)}\right] +$$
$$\cdots + \lambda^{R^*_\gamma}\left[\dfrac{1}{\mu_1(\mu_1+\gamma)\cdots\{\mu_1+(R^*_\gamma-1)\gamma\}}\right].$$

As $\mu_2 > \mu_1$, hence, Denominator of $L_I(\mu_1) >$ Denominator of $L_I(\mu_2)$. It leads to $L_I(\mu_1) < L_I(\mu_2)$ and $\gamma^*_2 < \gamma^*_1$.

We now consider L_A as a QoS measure. The proof for $\gamma_2^* > \gamma_1^*$ for $\mu_2 > \mu_1$ is similar to the QoS measure L_I. Now, let's understand the impact of λ on PSNE.

Let's say, $L_A(\lambda) = 1 - \dfrac{\lambda^{R_\gamma^*}}{P\,Q(\lambda)}$, where, $P = \prod_{k=0}^{R_\gamma^* - 1}(\mu + k\gamma)$ and $Q(\lambda) = 1 + \dfrac{1}{\mu}\sum_{j=1}^{R_\gamma^*}\dfrac{\lambda^j}{\prod_{i=1}^{j-1}(\mu + i\gamma)}$. Let's say, $L_B(\lambda) = \dfrac{\lambda^{R_\gamma^*}}{P Q(\lambda)}$.

If

$$\dfrac{\partial L_B(\lambda)}{\partial \lambda} = \dfrac{Q(\lambda)\,R_\gamma^*\,\lambda^{R_\gamma^* - 1} - \lambda^{R_\gamma^*}\dfrac{\partial Q(\lambda)}{\partial \lambda}}{(Q(\lambda))^2} < 0.$$

Then

$$Q(\lambda)\,R_\gamma^*\,\lambda^{R_\gamma^* - 1} < \lambda^{R_\gamma^*}\dfrac{\partial Q(\lambda)}{\partial \lambda}.$$

Let's divide both sides by $\lambda^{R_\gamma^* - 1}$, we get: $Q(\lambda)\,R_\gamma^* < \lambda\dfrac{\partial Q(\lambda)}{\partial \lambda}$. Here, $\dfrac{\partial Q(\lambda)}{\partial \lambda} = \dfrac{1}{\mu}\sum_{j=1}^{R_\gamma^*}\dfrac{j\lambda^{j-1}}{\prod_{i=1}^{j-1}(\mu + i\gamma)}$. After substituting the values of $Q(\lambda)$ and $\dfrac{\partial Q(\lambda)}{\partial \lambda}$, we get the following:

$$\dfrac{1}{\mu}\sum_{j=1}^{R_\gamma^*}\dfrac{(j - R_\gamma^*)\,\lambda^j}{\prod_{i=1}^{j-1}(\mu + i\gamma)} > R_\gamma^*.$$

Clearly, this condition cannot be true as $R_\gamma^* \geq 0$. Therefore, $\dfrac{\partial L_B(\lambda)}{\partial \lambda} > 0$. It leads to $\dfrac{\partial L_A(\lambda)}{\partial \lambda} < 0$, and hence, from Eq. (1), $\gamma_2^* > \gamma_1^*$.

References

1. Ayhan, H.: Optimal admission control in queues with abandonments. Oper. Res. Lett. **50**(6), 712–718 (2022). https://doi.org/10.1016/j.orl.2022.10.012
2. Benelli, M., Hassin, R.: Rational joining behavior in a queueing system with abandonments. Oper. Res. Lett. **49**(3), 426–430 (2021). https://doi.org/10.1016/j.orl.2021.04.004
3. Hemachandra, N., Patil, K., Tripathi, S.: Equilibrium points and equilibrium sets of some GI/M/1 queues. Queue. Syst. **96**(3), 245–284 (2020). https://doi.org/10.1007/s11134-020-09677-5
4. Hemachandra, N., Rajesh, K.S.N., Qavi, M.A.: A model for equilibrium in some service-provider user-set interactions. Ann. Oper. Res. **243**, 95–115 (2016). https://doi.org/10.1007/s10479-015-2037-8

5. Hyon, E., Jean-Marie, A.: Optimal control of admission in service in a queue with impatience and setup costs. Perform. Eval. **144**, 102134 (2020). https://doi.org/10.1016/j.peva.2020.102134
6. Koçağa, Y.L., Ward, A.R.: Admission control for a multi-server queue with abandonment. Queue. Syst. **65**, 275–323 (2010). https://doi.org/10.1007/s11134-010-9176-z
7. Rubinstein, A.: Comments on the interpretation of game theory. Econometrica J. Econometric Soc., 909–924 (1991). htttps://www.jstor.org/stable/2938166
8. Sinha, S.K., Rangaraj, N., Hemachandra, N.: Pricing surplus server capacity for mean waiting time sensitive customers. Euro. J. Oper. Res. **205**(1), 159–171 (2010). https://doi.org/10.1016/j.ejor.2009.12.023
9. Wolff, R.W.: Stochastic Modeling and the Theory of Queues. Prentice-Hall (1989)
10. Wu, R., Ayhan, H.: Optimal admission control in queues with multiple customer classes and abandonments. Queue. Syst. **109**(1), 6 (2025). https://doi.org/10.1007/s11134-024-09934-x

Optimizing Stealth Infections in an SI²R Model with Active-to-Sleep Dynamics

Mohamed Arnouss[1,2](✉) ⓘ, Willie Kouam[1] ⓘ, and Yezekael Hayel[1] ⓘ

[1] CERI/LIA, Avignon University, Avignon, France
{arnold.kouam-kounchou,yezekael.hayel}@univ-avignon.fr
[2] LMCSA, Hassan II University, Casablanca, Morocco
mohamed.arnouss@alumni.univ-avignon.fr

Abstract. Over the years, cybercriminals have refined their tools by developing increasingly stealthy malware. One effective approach involves temporarily putting malware into a dormant state to evade detection, then reactivating it at the right moment to relaunch the infection. Notable examples, such as *Emotet and Marap*, well illustrate this strategy: the former is capable of remaining inactive for several weeks before launching new attacks, while the latter passively gathers minimal system data while awaiting further instructions. These behaviors suggest the presence of two distinct categories of infected nodes: active and dormant. Motivated by these practical illustrations, we propose an SI²R model in which an infected node can transition into a dormant, undetectable state and be reactivated at a speed controlled by the attacker. The goal is to identify optimal transition strategies (both for dormancy and reactivation) to optimize the peak number of dormant nodes reached during the process, prepared to strike at the most effective moment. This model provides a mathematical framework for analyzing delayed cyber threats and the attacker's timing strategies while also informing proactive defense mechanisms against such latent threats (DISTRIBUTION A. Approved for public release: distribution unlimited).

Keywords: Malware · Optimization · Epidemic processes · Compartmental model

1 Introduction

The rapid evolution of cyberattacks has introduced complex challenges to cybersecurity, with stealth malware emerging as a particularly insidious threat due to its ability to evade detection through dormancy strategies [14]. Unlike traditional aggressive malware that risks early detection through immediate propagation, stealth malware strategically alternates between active and passive states to maximize persistence while minimizing detection probability. Emotet[1], for

---
[1] https://www.databreachtoday.com/emotet-botnet-shows-signs-revival-a-12964.

instance, is notorious for halting propagation to bypass antivirus scans, resuming its actions under favorable conditions, while Marap operates in passive data collection mode while awaiting remote instructions. These examples highlight a growing trend: modern malware prioritizes persistence and invisibility, increasing the potential for widespread damage in networked systems. The propagation of such malware through complex networks presents unique challenges, particularly in scale-free networks where hub nodes create vulnerabilities that can be exploited through delayed activation and stealth-based propagation tactics [5,8,13]. The heterogeneous nature of modern networks requires malware models that can capture both the structural network properties and the behavioral adaptability of stealth threats [7,11]. The analogy between biological epidemic models and computer virus propagation has been well-established since the foundational work of Kermack and McKendrick 1927 [4] on the SIR model. This biological paradigm provides a robust mathematical framework for understanding malware spread, as both biological pathogens and computer viruses exhibit similar propagation patterns through susceptible populations [2,6,9]. These models rely on fundamental concepts from mathematical epidemiology, including the basic reproduction number (R_0) and epidemic thresholds that determine virus-free equilibria [1]. Network-specific dynamics, particularly in scale-free networks, have been shown to significantly influence epidemic behavior and the effectiveness of containment strategies [8]. However, traditional epidemiological models face significant limitations when applied to stealth malware. Most existing frameworks assume uniform infection states [10]; this assumption fails to capture the strategic dormancy mechanisms employed by advanced malware, where infected nodes can dynamically transition between active propagation and passive surveillance states. The lack of active-to-passive transition modeling in standard SIR/SIRS frameworks renders them inadequate for representing the behavioral complexity of stealth malware and advanced persistent threats. To address these limitations, we study the (SI^2R) model, which incorporates dual infected states: active and passive infections. This framework extends traditional single-compartment infection models by recognizing that infected nodes can exhibit fundamentally different behaviors while remaining compromised. Infected nodes can transition to passive states when detection risk is high and reactivate when network conditions favor propagation or when coordinated attacks are planned. This behavioral duality reflects the sophisticated decision-making processes observed in real-world stealth malware. The central optimization objective in our study focuses on maximizing the proportion of passive (dormant) nodes at the epidemic peak, representing an attacker's strategic goal of creating a latent reservoir for coordinated future attacks. Closely related to this, we examine whether maximizing the passive node count also prolongs the time nodes remain in the active-passive cycle before detection, potentially revealing a shared optimal strategy for both reservoir size and stealth duration. This paper makes three key contributions: (1) a mathematical framework for optimizing stealth infections within the SI^{2}R model, incorporating strategic active-to-passive transitions; (2) analysis of optimal transition parameters that maximize passive node populations at epidemic peaks; and (3) actionable insights for developing proactive countermeasures against latent threats, supported by numerical validation and sensitivity analysis.

2   Model Description

This section introduces the problem formulation by describing the continuous-time propagation model. In the remainder of the paper, we use $m_s(t)$, $m_a(t)$, $m_p(t)$, and $m_r(t)$ to denote the proportions of susceptible, active, passive, and resistant (or protected) nodes at time t, respectively, all belonging to the interval $[0,1]$. Since these represent population fractions, they satisfy the normalization condition $m_s(t) + m_a(t) + m_p(t) + m_r(t) = 1$. In our model, the active node fraction $m_a(t)$ is defined as those that propagate the virus, while the passive node fraction $m_p(t)$ represents those that are undetectable (or dormant). For simplicity, we omit the explicit time dependency in the remainder of the document. The resulting SI^2R model (for "susceptible–active $\leftrightarrow$ passive–protected") describing the malware infection process is represented by the following set of differential equations, illustrated by the Fig. 1:

$$(\mathcal{S}): \begin{cases} \dot{m}_s = -\lambda m_s m_a, \\ \dot{m}_p = \alpha m_a - \beta m_p, \\ \dot{m}_a = \lambda m_s m_a + \beta m_p - \alpha m_a - \delta m_a, \\ \dot{m}_r = \delta m_a. \end{cases}$$

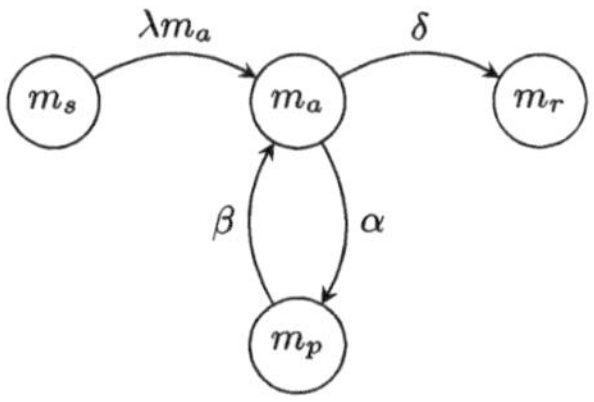

Fig. 1. System and compartmental diagram of the SI²R model

with parameters λ, α, β, $\delta \in [0,1]$ and initial conditions $m_s(0), m_a(0), m_p(0), m_r(0) \geqslant 0$, $m_s(0) + m_a(0) + m_p(0) + m_r(0) := 1$. In the system $\mathcal{S}$, λ is the malware's infection rate, that is, susceptible nodes that come into contact with an active infected node become infected at a rate λ. Upon infection, nodes transition either to the passive state at a speed α or to the resistant state with a rate δ, since infection-related activity may degrade system performance, increasing the likelihood that the computer owner detects and removes the virus. Once an active node becomes passive, it no longer propagates the virus; it is considered to be dormant and is therefore undetectable. However, these nodes can become active again for malware propagation with a speed β. As previously mentioned, the malware's objective is to identify the optimal transition pair (α, β) governing the active $\leftrightarrow$ passive switching of infected agents within the cycle in order to maximize the peak proportion of passive nodes reached during the process, as expressed in equation (1),

$$\max_{\alpha,\beta \in [0,1]} \left(\max_t m_p(t) \right) = \max_{\alpha,\beta \in [0,1]} G(\alpha, \beta), \qquad (1)$$

where $G(\alpha, \beta) = \max_t m_p(t)$ denotes the malware's objective function, which corresponds to the optimal peak proportion of passive agents attained during the process for a given pair (α, β). We also study the behavior of a node within the active $\leftrightarrow$ passive cycle; specifically, we investigate whether optimizing the total time nodes remain in the active $\leftrightarrow$ passive cycle before transitioning to the resistant state can lead to the *optimal peak in the fraction of passive nodes reached during the process*. A compromise emerges in the choice of α and β: when α tends to 0, active agents slowly transition to the passive state, thereby limiting the malware's ability to achieve its objective (since detection is more likely). Conversely, as α reaches 1, active agents quickly become passive, leaving too little time to propagate the infection effectively, which also restricts the overall impact. The same observation works for the parameter β.

3   Mathematical Analysis of the Model

We consider the assumption that the initial infection rate of susceptible devices ($\lambda m_s(0)$) is exactly the exit rate from the active state (δ), in such a way that neither propagation nor extinction dominates initially, in order to isolate and analyze precisely the behavior of the m_p peak without the risk of premature growth or decay. That is,

$$\lambda m_s(0) - \delta = 0. \tag{2}$$

The following proposition makes it possible to establish that the system under study is mathematically well-posed and significant by guaranteeing the existence, uniqueness, non-negativity, and conservation of the total population.

Proposition 1. *1. For any nonnegative initial parameters, there exists $\forall t \geqslant 0$ a unique solution $(m_s, m_a, m_p, m_r) \in C^1([0, \infty); \mathbb{R}^4)$ verifying $m_i(t) \geqslant 0$, $i \in \{s, a, p, r\}$; $m_s(t) + m_a(t) + m_p(t) + m_r(t) = 1$.*
2. No endemic equilibrium exists; the only feasible and general class of equilibria are the disease-free equilibria (DFE) *and the reproduction number is* $\mathcal{R}_* = \dfrac{\lambda m_s^*}{\delta}$, *at a given DFE $E_{s^*} = (m_s^*, 0, 0, 1 - m_s^*)$.*

Proof. 1. Define $u(t) = (m_s(t), m_a(t), m_p(t), m_r(t))^T$ and
$F(u) = \begin{pmatrix} -\lambda m_s m_a \\ \alpha m_a - \beta m_p \\ \lambda m_s m_a + \beta m_p - (\alpha + \delta) m_a \\ \delta m_a \end{pmatrix}$. Then system $(\mathcal{S})$ is rewritten as $u' = F(u)$ with $F \in C^1(\mathbb{R}^4)$ (containing polynomial entries). By the *Picard-Lindelöf* theorem, there is a unique maximal solution $u(t)$ on $[0, T_{\max})$. Summing the four equations: $\dfrac{d}{dt}(m_s + m_a + m_p + m_r) = -\lambda m_s m_p + (\alpha m_p - \beta m_a) + (\lambda m_s m_p + \beta m_a - (\alpha + \delta) m_p) + \delta m_p = 0$. Hence $N(t) = 1$ for all t and then $u(t)$ remains in the compact set $\{u : m_i \geqslant 0, \sum m_i = 1\}$; so no explosion occurs and $T_{\max} = \infty$.

2. - **Type of Equilibria:** From $\dot{m}_s = 0$, we have $m_s = 0$ or $m_a = 0$. If $m_a = 0$, by exploiting $\dot{m}_a = \dot{m}_p = 0$ we find $m_p = 0$. Then $m_s + m_r = 1$ with $m_s, m_r \geq 0$. Thus the set of DFEs: $E_{DFE}(s^*) = (m_s, m_a, m_p, m_r) = (m_s^*, 0, 0, 1 - m_s^*)$, $m_s^* \in [0,1]$. On the other hand, if $m_a > 0$; then $\dot{m}_s = 0$ implying $m_s = 0$, but $\dot{m}_r = \delta m_a > 0$ unless $m_a = 0$ which is absurd. Thus, no endemic state with $m_a > 0$ exists. Similarly, $m_p > 0$ alone is impossible; therefore, *no endemic equilibrium exists*, i.e., all equilibria are disease-free.

- **Reproduction Number $\mathcal{R}_*$:** The dynamics of the infected compartments of the system $\mathcal{S}$ can be rewritten as: $\dot{\mathbf{x}} = \mathcal{F}(\mathbf{x}) - \mathcal{V}(\mathbf{x})$, with $\mathbf{x} = \begin{bmatrix} m_p \\ m_a \end{bmatrix}$, $\mathcal{F}(\mathbf{x}) = \begin{bmatrix} 0 \\ \lambda m_s m_a \end{bmatrix}$, and $\mathcal{V}(\mathbf{x}) = \begin{bmatrix} \beta m_p - \alpha m_a \\ -\beta m_p + (\alpha + \delta) m_a \end{bmatrix}$. At the DFE $E_{s^*} = (m_s^*, 0, 0, 1 - m_s^*)$, the Jacobians are:

$$F = \left.\frac{\partial \mathcal{F}}{\partial(m_p, m_a)}\right|_{E_{s^*}} = \begin{bmatrix} 0 & 0 \\ 0 & \lambda m_s^* \end{bmatrix}, \quad V = \left.\frac{\partial \mathcal{V}}{\partial(m_p, m_a)}\right|_{E_{s^*}} = \begin{bmatrix} \beta & -\alpha \\ -\beta & \alpha + \delta \end{bmatrix}, \text{ and}$$

then

$$K = FV^{-1} = \begin{bmatrix} 0 & 0 \\ \frac{\lambda m_s^*}{\delta} & \frac{\lambda m_s^*}{\delta} \end{bmatrix}. \text{ The reproduction number is } \mathcal{R}_0 = \frac{\lambda m_s^*}{\delta}.$$

Under the above assumption 2, the following lemma holds. ∎

Lemma 1. *The function $f : t \mapsto f(t) = \dot{m}_p(t) = \alpha m_a(t) - \beta m_p(t)$, which is continuous on $[0, \infty)$ verifies $f(t) = 0 \implies f'(t) < 0$. In other words, the derivative m_p changes sign only once, with a clear crossing, which guarantees the uniqueness of the maximum of the function m_p.*

Proof. Let the function $f(t) = \dot{m}_p(t) = \alpha m_a(t) - \beta m_p(t)$, we have $f'(t) = \ddot{m}_p(t) = \alpha \dot{m}_a(t) - \beta \dot{m}_p(t)$. Consider $t_p \in [0, \infty)$ such that $f(t_p) = \dot{m}_p(t_p) = 0 \iff \alpha m_a(t_p) = \beta m_p(t_p)$, i.e., $m_p(t_p) = \frac{\alpha}{\beta} m_a(t_p)$. Moreover, $f'(t_p) = \alpha \dot{m}_a(t_p) - \beta \underbrace{\dot{m}_p(t_p)}_{=0} = \alpha \dot{m}_a(t_p)$. However,

$$\dot{m}_a(t_p) = \lambda m_s(t_p) m_a(t_p) + \beta \underbrace{\left(\frac{\alpha}{\beta} m_a(t_p)\right)}_{m_p(t_p)} - (\alpha + \delta) m_a(t_p)$$

$$= \lambda m_s(t_p) m_a(t_p) + \alpha m_a(t_p) - (\alpha + \delta) m_a(t_p) = m_a(t_p)\left(\lambda m_s(t_p) - \delta\right), \text{ i.e.,}$$

$$f'(t_p) = \alpha \dot{m}_a(t_p) = \alpha m_a(t_p)\left(\lambda m_s(t_p) - \delta\right).$$

Furthermore, the function $m_s : t \mapsto m_s(t)$ is strictly decreasing since $\dot{m}_s(t) = -\lambda m_s(t) m_a(t) \leq 0$ therefore, $m_s(t_p) < m_s(0)$. We deduce that $f'(t_p) < \alpha m_a(t_p)\left(\lambda m_s(0) - \delta\right) = 0$. In other words, at the point t_p when the function f reaches zero, its sign changes from positive to negative with a strictly negative slope. Moreover, a such maximum point is unique for the function m_p. Indeed, suppose by the absurd that there exists another time $t_2 > t_p$ such that $f(t_2) = 0$. Since $f(t_p) = 0$ and f decreases just after t_p, we have $f(t) < 0$ for t slightly greater than t_p. To reach another zero value at t_2, it would be necessary for f to go from negative to positive, that is to cross the zero a second time with a positive slope at this new crossing point. But as proven above, at any point where $f(t) = 0$, it must hold that $f'(t) < 0$. Therefore, a second crossing point t_2 if exist, will be a zero of f where the slope $f'(t_2)$ is positive, contradicting the fact that all t, zeros of f verify $f'(t) < 0$. In conclusion, there can be only one time t_p such that $f(t_p) = 0$, representing the unique maximum of the function m_p. ∎

Using the above lemma, we derive the following proposition, which asserts that the objective function is well-defined, meaning that the maximum of m_p exists for all $(\alpha, \beta) \in [0,1]^2$.

Proposition 2. *The objective function defines as:*

$$G(\alpha, \beta) = \begin{cases} m_p(0), & \text{if } \alpha\, m_a(0) - \beta\, m_p(0) < 0, \\ \dfrac{\alpha}{\beta} m_a(t_p^{\alpha,\beta}), & \text{otherwise}, \end{cases} \qquad (3)$$

where $t_p^{\alpha,\beta} \geqslant 0$ verifies $\dot{m}_p(t_p^{\alpha,\beta}) = 0$, is well-defined, and admits a unique maximum.

Proof. - If we suppose that $\alpha\, m_a(0) < \beta\, m_p(0)$, then $f(0) = \dot{m}_p(0) = \alpha\, m_a(0) - \beta\, m_p(0) < 0$, this means that from the initial time, the function $m_p : t \mapsto m_p(t)$ is decreasing. However, according to the above lemma 1, if $f(t_1) = 0$ at a given point $t_1 > 0$, then $f'(t_1) < 0$. If the function $f : t \mapsto f(t) = \dot{m}_p(t) = \alpha\, m_a(t) - \beta\, m_p(t)$ has to be zero at a point $t_p > 0$, it would first have to go from negative to positive, that is $f'(t_p) > 0$, which is absurd.

- If $\alpha\, m_a(0) \geqslant \beta\, m_p(0)$, then $\exists t_p^{\alpha,\beta} \in [0, \infty[$ such that $\dot{m}_p(t_p^{\alpha,\beta}) = 0$ since $\lim_{t \to \infty} m_p(t) = 0$; which implies $m_p(t_p^{\alpha,\beta}) = \dfrac{\alpha}{\beta} m_a(t_p^{\alpha,\beta})$. The uniqueness is proven by exploiting the Lemma 1. ∎

According to the Proposition 2 above, for fixed $(\alpha, \beta) \in [0,1]^2$ values, the function m_p admits a unique maximum. Since we aim at the end to determine the tuple (α, β) maximizing the function $G : (\alpha, \beta) \mapsto G(\alpha, \beta)$, we now have to demonstrate that G is a continuous function, i.e., $\max_{(\alpha,\beta) \in [0,1]^2} G(\alpha, \beta)$ exists.

Proposition 3. *Consider $\alpha, \beta \in [0,1]^2$, the function $G(\alpha, \beta)$ is continuous on $[0,1]^2$.*

Proof. We rewrite the system $(\mathcal{S})$ as $\dot{N}(t) = U(N, \alpha, \beta)$, where $N = \begin{bmatrix} m_s \\ m_p \\ m_a \\ m_r \end{bmatrix}$

and $U(N, \alpha, \beta) = \begin{bmatrix} -\lambda m_s m_a \\ \alpha m_a - \beta m_p \\ \lambda m_s m_a + \beta m_p - \alpha m_a - \delta m_a \\ \delta m_a \end{bmatrix}$. The vector $U(N, \alpha, \beta)$ is a polynomial function of N and (α, β), hence it is continuously differentiable

($\mathcal{C}^1$) in both parameters on $\mathbb{R}^4 \times [0,1]^2$. Furthermore, since the differential system is autonomous (i.e., does not explicitly depend on time t), by the Cauchy–Lipschitz (Picard–Lindelöf) theorem, the system with known initial conditions, admits a unique solution $N(t, \alpha, \beta)$ on some interval $[0, T]$. Moreover, by standard results on the continuous dependence of solutions with respect to parameters [3], the solution map $(t, \alpha, \beta) \mapsto M(t, \alpha, \beta)$ is continuous on $[0, T] \times [0, 1]^2$. Therefore, each component of this solution is therefore continuous in (t, α, β), that is $(t, \alpha, \beta) \mapsto m_p(t, \alpha, \beta)$ is continuous on $[0, T] \times [0, 1]^2$. Since the maximum of a continuous function over a compact interval depends continuously on parameters [12], the function G defined by $(\alpha, \beta) \mapsto G(\alpha, \beta)$ is continuous on $[0, 1]^2$. ∎

4 Behavior of an Infected Node and Approximation of m_p Function

4.1 Behavior of an Infected Node in the Active ↔ Passive Cycle

In the active state, each node has two possible transitions from active to passive or resistant, with transition probabilities respectively given by $P(A \to P) = \frac{\alpha}{\alpha + \delta}$, $P(A \to R) = \frac{\delta}{\alpha + \delta}$. Let X the random variable describing the number of times a node changes from active to passive before becoming resistant. X follows a geometric distribution $X \sim \text{Geometric}\left(\frac{\delta}{\alpha + \delta}\right)$, with a mathematical expectation given by, $\mathbb{E}[X] = \sum_{k=1}^{\infty} k P(X = k) = \frac{q}{p} = \frac{\alpha}{\delta}$ where $q = P(A \to P)$, $p = P(A \to R)$ and $P(X = k) = q^k \times p$ is the probability that the node will become passive k times before being detected and transition to resistant. Furthermore, let T_a and T_p the random variable representing the time a node spends in active and passive states respectively, then, $T_a \sim \text{Exponential}\left(\frac{1}{\alpha + \delta}\right)$ and $T_p \sim \text{Exponential}\left(\frac{1}{\beta}\right)$. The average time a node spends in one $A \leftrightarrow P$ cycle is given by, $T_c = \frac{1}{\alpha + \delta} + \frac{1}{\beta}$, then the expected total time a node spends in the $A \leftrightarrow P$ cycle before transitioning to the resistant state is expressed as, $T_c^*(\alpha, \beta) = \frac{\alpha}{\delta}\left(\frac{1}{\alpha + \delta} + \frac{1}{\beta}\right)$

Optimization of $T_c^*(\alpha, \beta)$: We analyze the function $T^*(\alpha, \beta)$ with respect to α and β.

$$\frac{\partial T_c^*}{\partial \alpha} = \frac{1}{\delta}\left(\frac{1}{\alpha + \delta} - \frac{\alpha}{(\alpha + \delta)^2} + \frac{1}{\beta}\right) = \frac{1}{\delta}\left(\frac{\alpha\delta + \delta^2}{(\alpha + \delta)^3} + \frac{1}{\beta}\right) \geqslant 0 \text{ and } \frac{\partial T_c^*}{\partial \beta} = -\frac{\alpha}{\delta\beta^2} < 0$$

That is, T_c^* decreases with β (slower transition from $P \to A$ increases in total cycle time) and T_c^* increases with α (accelerate transition speed from $A \to P$ increases in total cycle time).

Overall, to maximize the time a node spends in the $A \leftrightarrow P$ cycle, it is sufficient to *increase* α that means, *frequent returns to passive*, and to *decrease* β, i.e., *slower return to active*.

4.2 Approximation of m_p Function: Analysis of Optimal Parameters

The objective is to maximize the number of passive nodes at the epidemic peak (m_p) by controlling the parameters α and β. From the above system, we have the following relation $\dot{m}_a = -\dfrac{\dot{m}_s}{\lambda m_s}$ between $\dot{m}_s$ and $\dot{m}_a$. Now, substituting this into the equation for $\dot{m}_a$, we get:

$$\dot{m}_a = \lambda m_s \left(-\frac{\dot{m}_s}{\lambda m_s}\right) + \beta m_p - \alpha \left(-\frac{\dot{m}_s}{\lambda m_s}\right) - \delta \left(-\frac{\dot{m}_s}{\lambda m_s}\right)$$

Simplifying the terms, we obtain $\dot{m}_a = -\dot{m}_s + \beta m_p + \dfrac{\alpha \dot{m}_s}{\lambda m_s} + \dfrac{\delta \dot{m}_s}{\lambda m_s}$. Furthermore,

$$\frac{dm_a + dm_p}{dm_s} = -1 + \frac{\delta}{\lambda m_s} \iff dm_a + dm_p = -dm_s + \frac{\delta}{\lambda m_s} dm_s$$

$$\iff m_p = -m_s + \frac{\delta}{\lambda} \ln(m_s) + C - m_a, \text{ with } C = m_p(0) + m_s(0) - \frac{\delta}{\lambda} \ln(m_s(0)) + m_a(0), \text{ i.e.,}$$

$$\dot{m}_a = -\dot{m}_s + \beta \left(-m_s + \frac{\delta}{\lambda} \ln(m_s) + C - m_a\right) + \frac{\alpha \dot{m}_s}{\lambda m_s} + \frac{\delta \dot{m}_s}{\lambda m_s}$$

$$= -\dot{m}_s - \beta m_s + \frac{\beta \delta}{\lambda} \ln(m_s) + \beta C + \beta \frac{\dot{m}_s}{\lambda m_s} + \frac{\alpha \dot{m}_s}{\lambda m_s} + \frac{\delta \dot{m}_s}{\lambda m_s}$$

By exploiting the approximation for $\ln(m_s)$ coming from the Taylor expansion at order 1, i.e., $\ln(m_s) \approx \ln(m_s(0)) + \dfrac{1}{m_s(0)}(m_s - m_s(0))$, the equation for $\dot{m}_a$ can be rewritten as:

$$\dot{m}_a \approx -\dot{m}_s + \frac{\alpha \dot{m}_s}{\lambda m_s} + \frac{\delta \dot{m}_s}{\lambda m_s} - \beta m_s + \frac{\beta \delta}{\lambda}\left[\ln(m_s(0)) + \frac{1}{m_s(0)}(m_s - m_s(0))\right] + \beta C + \beta \frac{\dot{m}_s}{\lambda m_s}$$

$$\approx -\dot{m}_s + \left(\frac{\alpha}{\lambda m_s} + \frac{\delta}{\lambda m_s}\right)\dot{m}_s - \beta m_s + \frac{\beta \delta}{\lambda} \ln(m_s(0)) + \frac{\beta \delta}{\lambda m_s(0)}(m_s - m_s(0)) + \beta C + \beta \frac{\dot{m}_s}{\lambda m_s}$$

By hypothesis $\dfrac{\delta}{\lambda m_s(0)} = 1 \iff \delta = \lambda m_s(0)$, that is,

$$\frac{dm_a}{dt} \approx -\frac{dm_s}{dt} + \left(\frac{\alpha}{\lambda m_s} + \frac{\delta}{\lambda m_s}\right)\frac{dm_s}{dt} + \frac{\beta \delta}{\lambda} \ln(m_s(0)) - \frac{\beta \delta}{\lambda} + \beta C + \beta \frac{\frac{dm_s}{dt}}{\lambda m_s}$$

$$m_a \approx -m_s + \frac{\alpha + \delta + \beta}{\lambda} \ln m_s + \left(\frac{\beta \delta}{\lambda} \ln(m_s(0)) - \frac{\beta \delta}{\lambda} + \beta C\right) t + C_1$$

with $C_1 = m_a(0) + m_s(0) - \dfrac{\alpha + \beta + \delta}{\lambda} \ln(m_s(0))$. Now replacing this into the expression for m_p we get:

$$m_p \approx -m_s + \frac{\delta}{\lambda} \ln(m_s) + C - \left(-m_s + \frac{\alpha + \delta + \beta}{\lambda} \ln m_s + \left(\frac{\beta \delta}{\lambda} \ln(m_s(0)) - \frac{\beta \delta}{\lambda} + \beta C\right) t + C_1\right)$$

$$\approx -m_s + \frac{\delta}{\lambda} \ln(m_s) + C + m_s - \frac{\alpha + \delta + \beta}{\lambda} \ln m_s - \left(\frac{\beta \delta}{\lambda} \ln(m_s(0)) - \frac{\beta \delta}{\lambda} + \beta C\right) t - C_1$$

$$\approx \left(\frac{\delta}{\lambda} - \frac{\alpha + \delta + \beta}{\lambda}\right) \ln(m_s) + C - \left(\frac{\beta \delta}{\lambda} \ln(m_s(0)) - \frac{\beta \delta}{\lambda} + \beta C\right) t - C_1$$

Leveraging the differential equation $\dot{m}_s = -\lambda m_s m_a$ and replacing this expression into the expression of $\dot{m}_p$, which is set to zero at the passive nodes' epidemic peak, we obtain,

$$\left(\frac{-\alpha-\beta}{\lambda}\right)\frac{\dot{m}_s}{m_s} = \frac{\beta\delta}{\lambda}\ln(m_s(0)) - \frac{\beta\delta}{\lambda} + \beta C \iff \left(\frac{-\alpha-\beta}{\lambda}\right)\frac{-\lambda m_s m_a}{m_s} = \frac{\beta\delta}{\lambda}\ln(m_s(0)) - \frac{\beta\delta}{\lambda} + \beta C$$

$$\iff \left(\frac{-\alpha-\beta}{\lambda}\right)(-\lambda m_a) = \frac{\beta\delta}{\lambda}\ln(m_s(0)) - \frac{\beta\delta}{\lambda} + \beta C \iff m_a(t_p^{\alpha,\beta}) = \frac{\frac{\beta\delta}{\lambda}\ln(m_s(0)) - \frac{\beta\delta}{\lambda} + \beta C}{\alpha+\beta}$$

The epidemic peak proportion of passive nodes is, $m_p(t_p^{\alpha,\beta}) = \frac{\alpha}{\beta}m_a(t_p^{\alpha,\beta}) = \frac{\frac{\alpha\delta}{\lambda}\ln(m_s(0)) - \frac{\alpha\delta}{\lambda} + \alpha C}{\alpha+\beta}$

$$\frac{\partial m_p}{\partial \beta} = \frac{\partial}{\partial \beta}\left(\frac{\frac{\alpha\delta}{\lambda}\ln(m_s(0)) - \frac{\alpha\delta}{\lambda} + \alpha C}{\alpha+\beta}\right) = \frac{-\left(\frac{\alpha\delta}{\lambda}\ln(m_s(0)) - \frac{\alpha\delta}{\lambda} + \alpha C\right)}{(\alpha+\beta)^2} \quad \text{and}$$

$$\frac{\partial m_p}{\partial \alpha} = \frac{\partial}{\partial \alpha}\left(\frac{\frac{\alpha\delta}{\lambda}\ln(m_s(0)) - \frac{\alpha\delta}{\lambda} + \alpha C}{\alpha+\beta}\right) = \frac{\left(\frac{\delta}{\lambda}\ln(m_s(0)) - \frac{\delta}{\lambda} + C\right)\beta}{(\beta+\alpha)^2}.$$

Since $C = m_a(0) + m_p(0) + m_s(0) - \frac{\delta}{\lambda}\ln m_s(0)$, we have $\frac{\partial m_p}{\partial \alpha} \geqslant 0$ and $\frac{\partial m_a}{\partial \beta} < 0$, that is, α^* should be *the maximum possible value*, while β should be *the minimum possible value*.

In conclusion, taking into account the previous analysis, we conclude that to maximize the number of passive nodes at the end of the epidemic process (under the hypothesis 2), the attacker must rotate each infected node as much as possible in the loop $A \leftrightarrow P$.

5 Numerical Illustration

A numerical evaluation is conducted in this section to examine the influence of control parameters on the proposed computer virus propagation model. The experiments aim to achieve some main objectives: assessing the numerical error associated with the proposed approximation; validating, through simulations, the analytical results regarding the control parameters (α, β) that maximize the epidemic peak fraction of passive nodes by the end of the process; and demonstrating the existence of a non-trivial optimal pair of parameters (α, β), depending on the initial conditions.

5.1 Approximation Accuracy

To assess the accuracy of the proposed approximation, we numerically compare the approximated peak of the number of passive nodes with the exact peak obtained from the system $\mathcal{S}$. Our goal is to show that the approximation closely matches the exact peak, with only a small maximum absolute error. For the first simulation, we use the parameters $\alpha = 0.65$, $m_a(0) = 0.15$, $m_s(0) = 0.8$, $m_p(0) = 0.05$, $m_r(0) = 0$, and $\lambda = 0.2$. The result is presented in Fig. 2, where the proximity of the approximate and exact peak values is clearly illustrated.

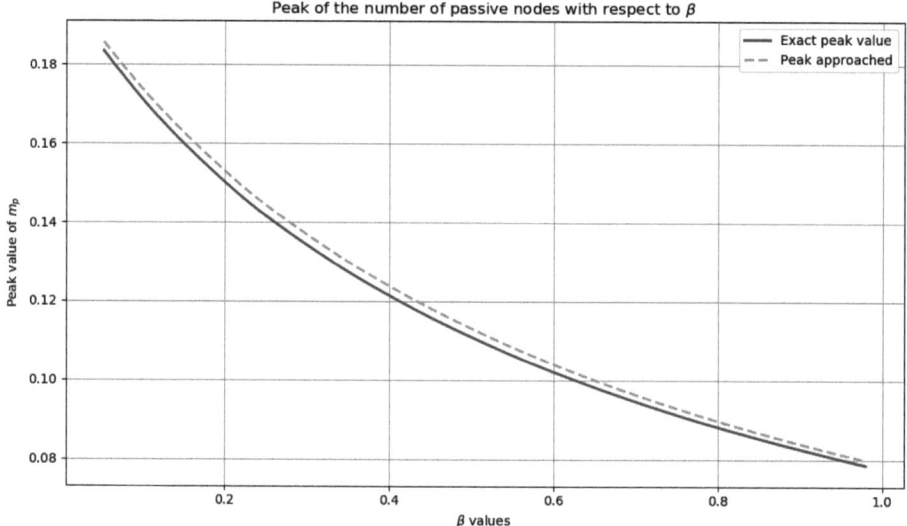

Fig. 2. peak proportion of passive nodes: *the curve of the approximate peaks follows well that of the exact peaks with a small error, indicating the validity of the proposed approximation*

To further evaluate the robustness of the approximation, we analyze the absolute error between the exact and approximate peak values of passive nodes across various settings. Specifically, we fix α and vary β values for each case. Table 1 summarizes the maximum absolute errors observed under these variations. In this second set of experiments, we use the initial conditions $m_a(0) = 0.01$, $m_s(0) = 0.99$, $m_p(0) = 0$, $m_r(0) = 0$, and set $\lambda = 0.2$. These results demonstrate that the proposed approximation consistently yields a low error, confirming its effectiveness for estimating the peak proportion of passive nodes in the studied system.

5.2 Optimal Parameters Analysis

In our model, the attacker strategically adjusts the control parameters α and β to maximize the peak proportion of passive nodes by the end of the epidemic propagation. Analytical results reveal that the peak fraction of passive

Table 1. Maximum error (%) = $\dfrac{|m_p(t_p^{\alpha,\beta}) - \tilde{m}_p(t_p^{\alpha,\beta})|}{m_p(t_p^{\alpha,\beta})} \times 100$, for each value of α by varying different values of the β parameter: *the maximum approximation error of the peak of the number of passive nodes at the end of the process does not exceed 2%. This shows that the proposed approximation is satisfactory. This approximation improves with increasing values of alpha.*

α values	0.15	0.22	0.3	0.37	0.44	0.51	0.59	0.66	0.73	0.8	0.88	0.95
Max Error (%)	2.06	1.21	0.792	0.5665	0.4307	0.3414	0.2780	0.2313	0.1977	0.1699	0.148	0.128

nodes reaches its maximum when α attains its highest value and β its lowest. Conversely, this suggests that any parameter pair deviating from this condition results in a reduced peak. To illustrate this, we consider the initial conditions $m_a(0) = 0.2$, $m_s(0) = 0.8$, $m_p(0) = 0$, $m_r(0) = 0$, and a transmission rate $\lambda = 0.7$, which corresponds to $\delta = 0.56$. We then vary α and β within the range $[0.15, 0.9]$.

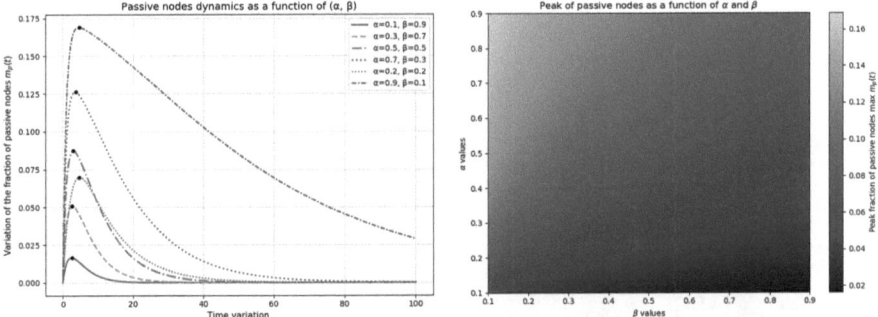

(a) Peak proportion variation for different pairs (α, β) including sub-optimal ones.

(b) Optimal pair (α, β) maximizing the peak proportion of passive nodes.

Fig. 3. Peak proportion of passive nodes depending of (α, β) pairs: *the results indicate that the peak proportion of passive nodes is maximized when α is maximal and β minimum. This behavior aligns with expectations, as deviations from these optimal conditions lead to a significant reduction in the peak. For instance, when $\alpha = 0.1$ and $\beta = 0.9$ the peak decreases by up to 90.21% highlighting the system's sensitivity to these parameters.*

Figure 3a shows the peak proportion of passive nodes when the optimal control parameters are not applied, that is, when alternative pairs of α and β are selected. As anticipated, the peak under these suboptimal parameters is consistently lower than the peak obtained with maximal α and minimal β. Furthermore, Fig. 3b presents a detailed color map depicting the peak fraction of passive nodes over 100 sampled values of α and β. This visualization confirms the analytical insight: the peak is indeed maximized when α is at its maximum and β

at its minimum. This finding highlights the critical role of parameter tuning in the attacker's strategy to optimize the epidemic's impact on passive nodes.

The mathematical analysis presented so far has been conducted under the assumption that the initial susceptible fraction satisfies the equality $\lambda m_s(0) = \delta$. Under this condition, the attacker's optimal strategy consists of maximizing the transition speed α and minimizing β in order to maximize the peak proportion of passive nodes. However, when this assumption is relaxed (especially when $\lambda m_s(0) > \delta$), the optimal values of α and β are no longer necessarily found at their boundary values. Instead, they may lie within the interior of the parameter space. This behavior is illustrated in Fig. 4, where we consider the initial state parameters: $m_s(0) = 0.88$, $m_a(0) = 0.01$, $m_p(0) = 0.11$, $m_r(0) = 0$, with $\lambda = 0.01$ and $\delta = 0.005$.

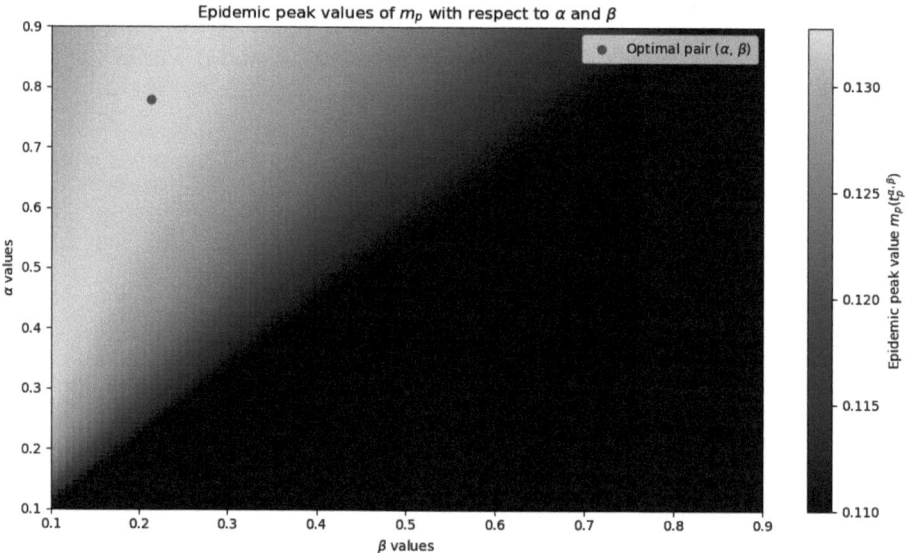

Fig. 4. Illustration of a non-trivial optimal pair (α, β) that maximizes the peak number of passive nodes at the end of the process: *for the given parameters, the optimal pair is $(\alpha, \beta) = (0.779, 0.213)$, yielding a final passive node fraction of 0.1327. This highlights that our model admits non-trivial solutions depending on the initial conditions, thereby revealing richer dynamics beyond simple boundary strategies.*

6 Conclusion

In this work, we investigate the spread of malware across a digital network by employing a novel variant of the Susceptible-Infected-Infected-Resistant (SI^2R)

epidemic model. Unlike classical approaches in the literature, our model introduces a mechanism whereby active (infectious) nodes may transition into a passive (dormant) state at a controllable speed α. In this state, nodes become undetectable, effectively evading security measures while halting direct propagation. However, these passive nodes retain the potential to be reactivated at a speed β, re-entering the active state and pursuing the spread of malware. The central objective of our analysis is to maximize the peak fraction of passive nodes during the propagation process. We demonstrate that under a specific assumption, this maximum is achieved when the transition speed from active to the passive state (α) is *as high as possible*, while the reactivation rate (β) is *minimized*. We have shown that, this strategy prolongs the time nodes spend cycling between passive and active states before eventual removal (resistance), thereby sustaining the infection within the network for a longer period. Interestingly, we also uncover scenarios in which the optimal pair (α, β) does not lie on the boundary of the parameters space, indicating a non-trivial trade-off between concealment and propagation, highlighting the richness and flexibility of our framework. Future directions include relaxing some assumptions and deriving a full mathematical characterization of the optimal parameters when they lie in the interior of the domain. Additionally, we aim to explore the impact of time-varying parameters $\alpha(t)$ and $\beta(t)$ on the epidemic dynamics and control strategies.

Acknowledgements. Research was sponsored by the Army and was accomplished under Grant Number W911NF-24-0149. The views and conclusions contained in this document are those of the authors and should not be interpreted as representing the official policies, either expressed or implied, of the U.S. Army or the U.S. Government. The U.S. Government is authorized to reproduce and distribute reprints for Government purposes notwithstanding any copyright notation herein.

References

1. Diekmann, O., Heesterbeek, J.A.P.: Mathematical Epidemiology of Infectious Diseases. Wiley, Hoboken (2000)
2. Gelenbe, E.: Dealing with software viruses: a biological paradigm. Inf. Secur. Technol. Rep. **12**(4), 242–250 (2007)
3. Hartman, P.: Ordinary differential equations. SIAM (2002)
4. Kermack, W.O., McKendrick, A.G.: A contribution to the mathematical theory of epidemics. Proc. Roy. Soc. A **115**(772), 700–721 (1927)
5. Li, W., Wang, Y., et al.: Modeling propagation dynamics of stealth malware. IEEE Access **7**, 88792–88804 (2019)
6. Mishra, B.K., Saini, D.: Mathematical models on computer viruses. Appl. Math. Comput. **187**(2), 929–936 (2007)
7. Moreno, Y., Pastor-Satorras, R., Vespignani, A.: Dynamics of rumor spreading in complex networks. Phys. Rev. E **65**(3), 036130 (2002)
8. Pastor-Satorras, R., Vespignani, A.: Epidemic spreading in scale-free networks. Phys. Rev. Lett. **86**(14), 3200–3203 (2001)
9. Piqueira, J.R.C., et al.: Dynamic models for computer virus propagation. Math. Probl. Eng. (2008). https://doi.org/10.1155/2008/940526

10. Preciado, V.M., et al.: Optimal resource allocation for network protection against spreading processes. IEEE Trans. Control Netw. Syst. **1**(1), 99–108 (2013)
11. Rudd, E., Hannay, P., Rashid, A.: A survey of stealth malware: attacks, mitigation measures, and steps toward autonomous open world solutions. ACM Comput. Surv. **49**(4), 1–42 (2016)
12. Rudin, W.: Principles of mathematical analysis (2021)
13. Yadav, T., Selvakumar, S.: Detection of application layer DDOS attack by feature learning using stochastic feature transformation. Comput. Secur. **55**, 161–174 (2015)
14. Zhang, J., et al.: Polymorphic and metamorphic malware detection. J. Cybersecur. **5**(1), 1–12 (2019)

Optimizing Energy in Supervised Learning with Data Summarization: A Comparative Study

O. Haddaji[✉], O. Brun, and B.J. Prabhu

LAAS-CNRS, Université de Toulouse, CNRS, INSA, Toulouse, France
{ohaddaji,brun,bjprabhu}@laas.fr

Abstract. The significant computational demands of modern machine learning (ML) models raise growing concerns about their environmental impact and economic cost during training. To address this, data summarization techniques, which involve selecting a smaller, representative subset of the training data, offer a promising avenue for optimizing energy consumption without compromising model performance. This article investigates the influence of various data summarization algorithms—specifically random sampling, Facility Location (FL), and CRAIG—on three critical aspects of supervised learning (SL) training: model accuracy, energy consumption, and overall efficiency, which we quantify as the ratio of accuracy to energy consumption.

Our extensive experimental findings reveal that while sophisticated methods like FL and CRAIG aim to select highly representative subsets, random sampling consistently achieved a robust balance of accuracy and efficiency. This is particularly evident when accounting for the often significant pre-processing energy overheads incurred by more complex selection strategies. Furthermore, we emphasize the pivotal role of early stopping criteria in optimizing the overall energy efficiency of the training process. Our analysis demonstrates that strategic adjustments to these criteria can substantially reduce the number of training epochs required for convergence, thereby mitigating considerable energy waste.

Keywords: Data summarization · Energy-efficiency · Supervised learning

1 Introduction

The rapid advancement of machine learning (ML) has led to the development of increasingly complex models, many of which demand substantial computational resources for training. As a result, the energy consumption associated with ML training has become a growing concern, both in terms of environmental impact and economic cost, drawing significant attention from global organizations [7]. Addressing this challenge requires innovative strategies to optimize energy consumption without compromising model performance.

To reduce the energy consumption of machine learning training, several strategies have been explored. One common approach focuses on optimizing model architecture, which includes techniques like dropout and neural network pruning [4,6,12] to reduce computational complexity. Other lines of research focus on algorithmic and hardware optimizations, such as utilizing low-precision arithmetic [13] or specialized hardware like FPGAs and ASICs, to perform computations more efficiently. Our research, however, investigates a different, yet complementary, approach: reducing the size of the training data itself.

Data summarization techniques—selecting a smaller, more manageable subset of the training data—offer a promising path forward. They can take many forms, ranging from random sampling, which is easy to implement but may result in non-representative subsets, to informed selection strategies that aim to preserve the underlying data distribution and diversity. Representative subsets, often the solution of framing subset selection as a submodular maximization problem, are expected to be particularly effective in this regard. Data summarization techniques, such as submodular approximation [10,14,15], have been explored to accelerate machine learning training while maintaining accuracy.

Prior work on data summarization focuses on training time, and no attempt was done to quantify the impact on the energy consumption of training phase. Our research directly addresses this gap by rigorously evaluating how data summarization techniques influence the energy-accuracy tradeoff during model training, offering a novel contribution to the field.

In this article, we investigate the impact of various data summarization algorithms on three key aspects of supervised learning: model accuracy, energy consumption, and efficiency (quantified as the ratio of accuracy to energy consumption). We explore how different algorithms influence these metrics, reducing the energy consumption while maintaining desired model performance. We expect the insights and relationships derived from this study can be useful for researchers and practitioners looking to determine the most effective data summarization strategy for their specific optimization problems, such as minimizing energy subject to accuracy requirements or maximizing accuracy under an energy budget.

This paper is organized as follows. Section 2 details the representative sampling techniques investigated. Our experimental methodology, including the datasets and model architectures used, is outlined in Sect. 3 The experimental results and main findings on energy consumption, accuracy, and overall efficiency are presented in Sect. 4. Finally, Sect. 5 summarizes the key conclusions drawn from our research and presents the future directions.

2 Background Material on Representative Sampling

2.1 Facility Location Based Sampling

Energy consumption typically scales with training data size. Therefore, reducing the number of training examples—without significantly compromising performance—can lead to substantial energy savings. A promising strategy is to select a representative subset of the dataset that captures its overall diversity and structure [10,14,15].

The selection process is framed as an optimization problem, where the goal is to choose a subset of data points that best "covers" or represents the entire dataset. One well-established approach for this is the Facility Location Problem (FLP) [8], a classical optimization problem in operations research.

Formally, given a dataset $\mathcal{X} = \{x_1, x_2, \ldots, x_n\}$ and a similarity function $s : \mathcal{X} \times \mathcal{X} \to \mathbb{R}_{\geq 0}$, the Facility Location function $f : 2^{\mathcal{X}} \to \mathbb{R}$ is defined as:

$$f(S) = \sum_{x_i \in \mathcal{X}} \max_{x_j \in S} s(x_i, x_j) \tag{1}$$

where $S \subseteq \mathcal{X}$ is the selected subset, and $s(x_i, x_j)$ measures the similarity between data points x_i and x_j. The objective is to select a subset S of fixed size k that maximizes $f(S)$:

$$S^* = \arg \max_{S \subseteq \mathcal{X}, |S|=k} f(S)$$

The Facility Location function is submodular [5], exhibiting diminishing returns. This means the incremental value of adding an element decreases as the set grows. Mathematically, for any sets S and T such that $S \subseteq T$ and element $v \notin T$, the marginal gain $f(v \mid S)$ is greater than or equal to $f(v \mid T)$, where $f(v \mid S) = f(S \cup \{v\}) - f(S)$.

The FLP is NP-hard; however, thanks to its submodular structure, we can design efficient approximation algorithms using greedy approaches. Specifically, the Standard Greedy Algorithm iteratively selects the element yielding the highest marginal gain at each step, guaranteeing a $(1-1/e)$-approximation for monotone submodular functions [11]. Improvements like Lazy Greedy enhance efficiency by using a priority queue to reduce marginal gain evaluations [9].

Despite the theoretical promise, a primary challenge we face is the difficulty in accurately quantifying the similarity between images. Key techniques include Mean Squared Error (MSE) for pixel-wise differences, Normalized Correlation (NC) for linear relationships between intensity values, Mutual Information (MI) to measure statistical dependence, Cosine Similarity for focusing on orientation, and Euclidean Distance for calculating direct pixel value distance. For simplicity, we opted for Euclidean distance and converted it into a similarity score. Alternatively, one could measure the distance between data point gradients as explained below.

2.2 Gradient-Based Sampling: CRAIG

Given a training dataset $\mathcal{X} = \{x_1, x_2, \ldots, x_n\}$ and a loss function $f(w, x_i)$ for each data point x_i, the CRAIG method, developed by Mirzasoleiman et al. [10], selects a weighted subset $S \subseteq \mathcal{X}$ that minimizes the error in approximating the full gradient. The optimization problem can be formulated as a FLP (1) where

$s(x_i, x_j) = \max_{u,v} d(x_u, x_v) - d(x_i, x_j)$, and $d_{x_i, x_j} = \|\nabla f_{x_i}(w) - \nabla f_{x_j}(w)\|$ is the distance between the gradients at w for data points x_i and x_j.

Calculating the exact gradient of the loss function in deep neural networks (DNNs), given their vast number of parameters, is computationally intensive. To address this, the last-layer approximation can be used instead of exact gradients, as suggested in [3,10]. In the original CRAIG method, a representative subset of data points is selected at each epoch. In our experiments, to reduce the energy consumption, this selection was done only once after the initial gradients were computed.

3 Experimental Setup

Energy measurements were taken on the Grid5000 platform [1], a large-scale testbed. We specifically used identical configurations on the Rennes cluster (paradoxe-[1,3-4,12,21]), which has 5 nodes, 10 CPUs, and 260 cores in total. We monitored energy consumption at the operating system level using Mojitos [2], an open-source tool. Due to limited access to this specialized hardware, we measured the energy consumed per epoch across many different subset sizes. We assumed that the energy use per epoch remained constant.

We used three well-known benchmark datasets: MNIST (60,000 training, 10,000 test samples), Fashion MNIST (similar splits), and CIFAR-10 (50,000 training, 10,000 test samples). These datasets allowed us to test a range of complexities, from simple digit recognition to more complex object classification.

For our models, we used two types of Convolutional Neural Networks (CNNs). For MNIST and Fashion MNIST, we used a lightweight CNN with two convolutional layers, ReLU activation, and a fully connected layer. For the more complex CIFAR-10, we used a deeper CNN with three convolutional layers, batch normalization, dropout, and fully connected layers. All models were optimized using Adam, and cross-entropy loss was used as the loss function.

We compared three data summarization approaches: stratified random sampling, CRAIG and Facility Location Sampling (FL). Given that our datasets are pre-divided into classes, stratified random sampling operates by drawing a random sample from each class such that the proportion of each class in the sample is equivalent to its proportion in the dataset, thereby ensuring representative distribution. To ensure a fair comparison, both CRAIG and FL are also applied per class to select representative subsets from each class.

4 Energy and Efficiency in Data Subset Training

We measured the energy consumption associated with training machine learning models on CIFAR10 dataset. Our approach involved two main steps. First, we conducted 100 trials to estimate the energy usage for a single training epoch on randomly chosen data subsets. Our results showed a clear trend: the energy consumption per epoch rose significantly as the size of the data subset grew, specifically from 1.46 kJ for a 10% data subset to 7.6 kJ for a 75% data subset.

We will use these exact numerical per-epoch energy results, combined with the total number of epochs required for convergence, to calculate the total energy consumed per training run later in this article.

Next, we focused on determining the achievable accuracy and the total number of epochs to converge for various subset sizes, comparing two distinct algorithms: random sampling and facility location. As expected, convergence heavily depends on the chosen early stopping criterion, which directly impacts total energy consumption—a stricter criterion can lead to faster convergence (less energy) or slower convergence (more energy). In our experiments, training was terminated if the test accuracy didn't improve by at least 0.5% (tolerance of 0.005) for 5 consecutive epochs (patience of 5), or if a maximum limit of 500 epochs was reached.

We first present the results for the training phase for both random sampling as well as facility location techniques. We will comment on the pre-processing overheads for FL later on. Figure 1 shows the accuracy and the statistics of the number of epochs for convergence on the CIFAR-10 dataset for random sampling and FL. A plateau appears before the 500th epoch, indicating that the early stopping mechanism worked effectively.

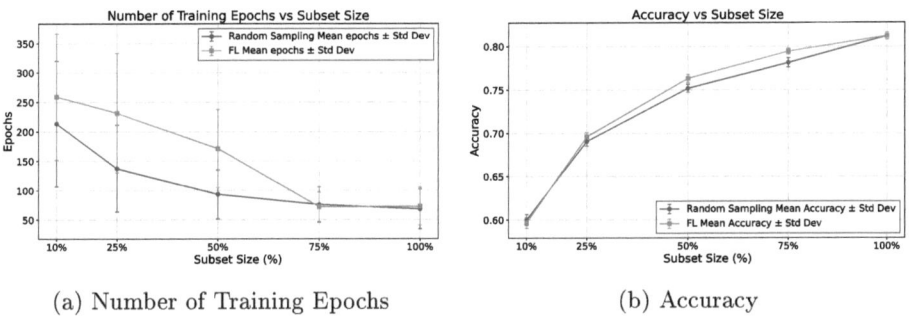

(a) Number of Training Epochs (b) Accuracy

Fig. 1. Statistics of the number of training epochs and accuracy vs subset size.

In order to understand why convergence may take much longer with smaller subsets, we look at the sample path of accuracy from one experiment in Fig. 2 for random sampling (left) and FL (right). The validation accuracy has higher fluctuations for smaller subset sizes which makes it more difficult to quickly meet the condition of 0.5% tolerance in the early stopping criterion. For example, for 1%, the accuracy plateaus by epoch 50 but the early stopping criterion is met only after epoch 350. For nearly 300 epochs, the training is being performed and energy is being consumed without actually gaining too much in accuracy. This could be one avenue where energy savings could be made by defining a slightly looser early stopping criterion.

4.1 Facility Location vs. Random Sampling

We observed that FL offers no advantage over random sampling in terms of training epochs to convergence. In fact, our results in Fig. 2a indicate that FL selected subsets, on average, actually require more training epochs. This can be explained by the selection bias of similarity-based methods. While they favor 'easy' central examples, they are less likely to select the 'hard' or boundary-case examples crucial for refining a model's decision boundaries. Random sampling, by contrast, naturally includes these challenging points, which often speeds up convergence.

Convergence variability is due to mini-batch shuffling, which creates different gradient updates and learning trajectories in each epoch. However, as Fig. 2b shows, the similarity-selected subsets still consistently achieved higher accuracy than random sampling, despite requiring more epochs.

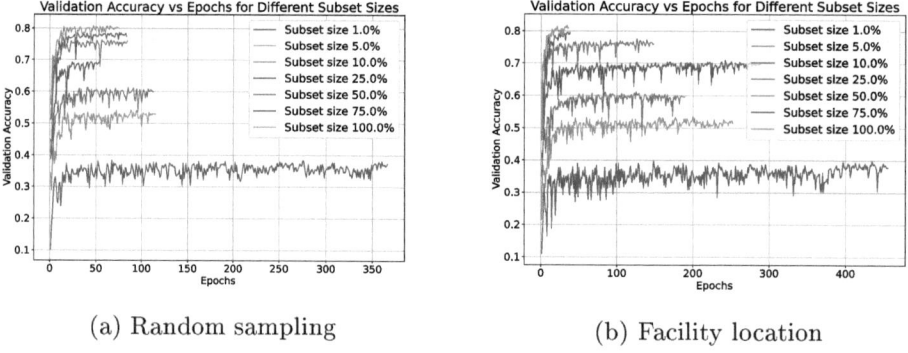

Fig. 2. Evolution of accuracy over training epochs for different subset sizes. CIFAR-10 dataset.

The observation that FL provides only marginal accuracy improvements over random sampling is likely due to the lack of a naturally suitable similarity function for the datasets. In our experiments, this function was taken as an inverse to euclidean distance (higher similarity for closer points) but this choice is arbitrary. Further, once the datasets have been subdivided into classes, the data inside each class seems to be sufficiently homogeneous so that random sampling is not too far off from the FL when we compare the values of the submodular objective function, O, given in (1) evaluated on the respective subsets. Table 1 shows $O(S)/O(F)$ averaged over 100 experiments for the two algorithms, where F is full the dataset and S is the subset provided by the algorithm. Although the subset chosen by FL does have a better value uniformly over subset sizes, the gains start becoming interesting only for very small subset sizes (~ 10). Note that in this comparison the objective function is very much favorable to FL sampling technique and it is not straightforward to predict the gain (or loss) in accuracy that can be expected from a given gain in the objective value since the choice of the objective function was somewhat arbitrary.

Table 1. Fraction of FL score to the full dataset. CIFAR-10 dataset

$SubsetSize/Class$	10	100	500	750	1250	2500
FL		0.94	0.95	0.96	0.96	0.97	0.98
Random		0.91	0.94	0.95	0.95	0.96	0.97

4.2 Pre-processing Energy Overheads

The marginal accuracy advantage comes at an additional cost: the pre-processing energy overheads associated with the subset selection itself. We measured this energy on Grid'5000 using various subset sizes from the CIFAR10 dataset, repeating each experiment 10 times. This selection energy is crucial, as it must be added to the training energy (which is calculated by multiplying the number of epochs by the energy consumed per epoch) to determine the total energy consumption for each complete training run.

Figure 3 illustrates energy consumption for FL and CRAIG. FL consistently consumed around 5 kJ across different subset sizes. In contrast, CRAIG showed higher energy consumption, ranging from 6.46 kJ to 8.12 kJ for selecting 1% to 75% of the data. In fact, to calculate the gradient using the last layer approximation method, a forward pass is needed. A forward pass is the process of feeding input data through a neural network to produce an output prediction. It involves a series of matrix multiplications and activation function applications across the network's layers, moving from the input layer towards the output layer. Consequently, the energy required to obtain a subset using CRAIG depends on the complexity of the model.

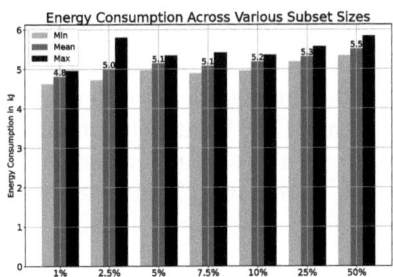
(a) Computational Energy for FL

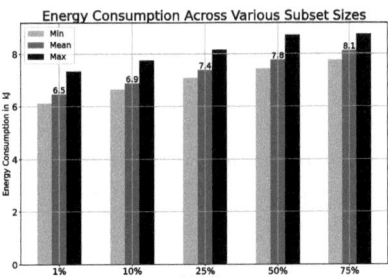
(b) Computational Energy for CRAIG

Fig. 3. Energy consumption for subset selection using facility location and CRAIG on CIFAR10 datasets. Each graph shows the minimum, mean, and maximum energy consumption for different subset sizes.

4.3 Comparing Efficiency

We define efficiency as accuracy divided by the mean total energy consumption. To evaluate the efficiency of different sampling methods, we conducted experiments on both CIFAR-10 and MNIST datasets, plotting smoothed fitted functions based on experimental data.

As shown in Fig. 4, the efficiency trends vary between datasets. For CIFAR-10 (Fig. 4a), random sampling consistently demonstrated higher efficiency than FL across most subset sizes. The only exception is at a 75% subset size, where their efficiencies become comparable. The efficiency for CIFAR-10 typically falls between 0.1 and 0.2.

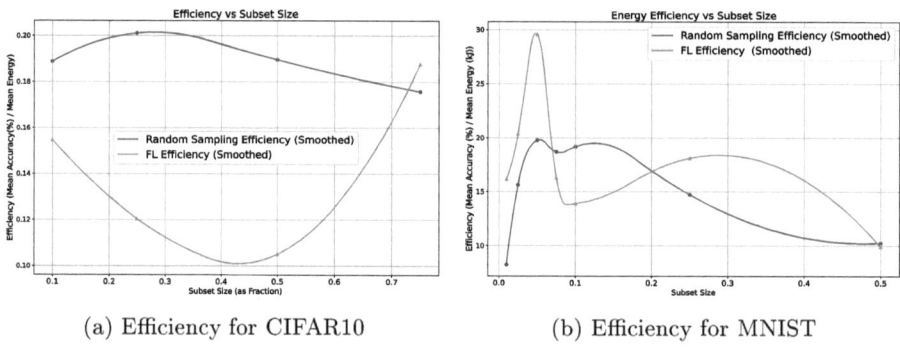

(a) Efficiency for CIFAR10 (b) Efficiency for MNIST

Fig. 4. Efficiency curves for different algorithms. CIFAR10 and MNIST datasets.

In contrast, for MNIST (Fig. 4b), the efficiency curves for random sampling and FL are more alike. Random sampling shows higher efficiency for subset sizes between 7.5% and 20%, but FL generally maintains a higher overall efficiency. The efficiency for MNIST ranges between 5 and 30. This significant difference is easily explained by the inherent characteristics of the MNIST dataset: it requires less data to achieve maximum accuracy and utilizes a less complex model for training, consequently demanding less energy. This observation lends credibility to our defined efficiency criteria.

4.4 Impact of Stopping Criterion

Our analysis of training and test times reveals a crucial insight regarding computational efficiency: the testing phase frequently demands significantly more time than training, especially when using small data subsets. This phenomenon arises because evaluation necessitates processing the entire fixed test set of 10,000 samples after each epoch to check the stopping criteria.

To illustrate with a simple calculation: if the 10,000-sample test set represents 20% of the training set size, and we perform a test after each of 100 epochs, this adds a substantial computational overhead. Considering that training involves both a forward and a backward pass, while testing primarily requires a forward pass, evaluating on 20% of the training data effectively adds $100 \times (0.5 \times 20\%) = 10$ times the effort of training on the full training set. Consequently, if we use a training data subset of 10% of the original size, the energy consumed by repeated testing can equal that of the training itself.

An alternative approach to mitigate this repeated testing after each epoch is to train for a predetermined number of rounds and evaluate only once at the end of the training. However, given that we do not know the required number of training epochs in advance, a practical solution to this issue is to modify the stopping criteria to be based on training accuracy rather than test accuracy, thereby allowing test accuracy to be measured only once at the very end of the training process.

To further investigate this and enable a more effective comparison between CRAIG and FL, we changed the stopping criteria to be applied to the training accuracy. Additionally, we introduced a new stopping criterion: training was terminated if the test accuracy did not improve by at least 1% (a tolerance of 0.01) for 3 consecutive epochs (a patience of 3), or if a maximum limit of 500 epochs was reached. This adjustment aimed to reduce the number of training epochs and thus improve overall efficiency.

Figure 5 compares the efficiency of two CRAIG application strategies against FL on the MNIST dataset, all evaluated under this new criterion. The red curve represents the efficiency when CRAIG is applied only once at the beginning of training. In this scenario, the overhead energy of applying CRAIG is simply added as a fixed cost to the total training energy for each subset size. Conversely, the green curve illustrates the efficiency when CRAIG is applied every five epochs, as suggested in the CRAIG paper [10]. For this approach, the overhead energy of applying CRAIG is dynamically scaled by a coefficient of "total number of epochs/5 + 1" for each subset size. This increased energy overhead results in lower efficiency compared to the single application method.

The results clearly show that FL is considerably more efficient than both CRAIG application methods. Notably, by reducing the number of training

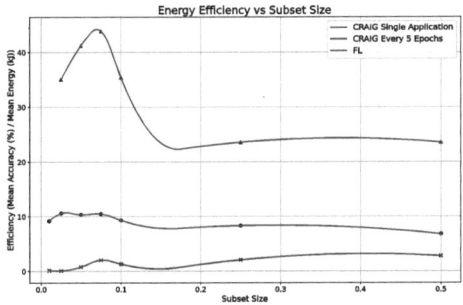

Fig. 5. Comparing the efficiency of CRAIG and FL using MNIST dataset

epochs, this new criterion significantly enhances FL's overall efficiency, outperforming the results previously observed for MNIST in Fig. 4b.

5 Conclusions

This study explored the impact of various data summarization techniques—random sampling, Facility Location (FL), and CRAIG—on the energy consumption and accuracy of supervised learning models.

Random sampling consistently demonstrated a strong balance of accuracy and efficiency, particularly when accounting for the pre-processing energy overheads associated with the other methods. Although FL can select more representative subsets, it occasionally necessitated more training epochs than random sampling, and its marginal accuracy improvements often did not justify the additional computational cost. CRAIG proved to be significantly less efficient due to its substantially higher pre-processing energy demands.

Our research also highlighted the critical role of early stopping criteria in optimizing overall efficiency. Adjusting these criteria can substantially reduce training epochs and mitigate energy waste.

Ultimately, while sophisticated summarization methods exist, the simpler approach of random sampling frequently emerges as a highly competitive and energy-efficient choice for training supervised learning models. Our findings, however, are based on a limited set of benchmark image datasets and CNN architectures, and future work is needed to validate these conclusions on a broader range of data types and larger, real-world datasets. The choice of similarity function for FL was also arbitrary, and exploring alternative functions may yield different results. Future studies could also investigate the "doubling trick" as an alternative to our proposed training-based stopping criterion to balance energy savings with the use of a test set.

Acknowledgement. This work was partially supported by the French National Research Agency (ANR) under grant ANR-22-CE23-0024 (project DELIGHT).

References

1. Grid'5000 – large-scale flexible experimental testbed. https://www.grid5000.fr. Accessed 18 Oct 2025
2. MoJITOs – monitoring java infrastructure for tracing and observability solutions (2023). https://gitlab.irit.fr/sepia-pub/mojitos
3. Ash, J.T., Zhang, C., Krishnamurthy, A., Langford, J., Agarwal, A.: Deep batch active learning by diverse, uncertain gradient lower bounds. In: International Conference on Learning Representations (2020). https://openreview.net/forum?id=ryghZJBKPS
4. Frankle, J., Dziugaite, G.K., Roy, D.M., Carbin, M.: Linear mode connectivity and the lottery ticket hypothesis. In: Proceedings of the 37th International Conference on Machine Learning. ICML'20, JMLR.org (2020)

5. Fujishige, S.: Submodular Functions and Optimization. Elsevier, Amsterdam (2005)
6. Han, S., Pool, J., Tran, J., Dally, W.J.: Learning both weights and connections for efficient neural networks. In: Proceedings of the 29th International Conference on Neural Information Processing Systems - Volume 1. NIPS'15, pp. 1135–1143. MIT Press, Cambridge, MA, USA (2015)
7. IEA: Energy and AI (2025). https://www.iea.org/reports/energy-and-ai
8. Kuehn, A.A., Hamburger, M.J.: A heuristic program for locating warehouses. Manag. Sci. **9**(4), 643–666 (1963)
9. Minoux, M.: Accelerated greedy algorithms for maximizing submodular set functions. In: Stoer, J. (ed.) Optimization Techniques. LNCIS, vol. 7, pp. 234–243. Springer, Heidelberg (1978). https://doi.org/10.1007/BFb0006528
10. Mirzasoleiman, B., Bilmes, J., Leskovec, J.: Coresets for data-efficient training of machine learning models. In: Proceedings of the 37th International Conference on Machine Learning. ICML'20, JMLR.org (2020)
11. Nemhauser, G.L., Wolsey, L.A., Fisher, M.L.: An analysis of approximations for maximizing submodular set functions-i. Math. Program. 265–294 (1978)
12. Renda, A., Frankle, J., Carbin, M.: Comparing rewinding and fine-tuning in neural network pruning. In: International Conference on Learning Representations (2020). https://openreview.net/forum?id=S1gSj0NKvB
13. Sun, X., et al.: Hybrid 8-Bit Floating Point (HFP8) Training and Inference for Deep Neural Networks. Curran Associates Inc., Red Hook (2019)
14. Tschiatschek, S., Iyer, R., Wei, H., Bilmes, J.: Learning mixtures of submodular functions for image collection summarization. In: Proceedings of the 28th International Conference on Neural Information Processing Systems - Volume 1. NIPS'14. pp. 1413–1421. MIT Press, Cambridge, MA, USA (2014)
15. Wei, K., Iyer, R., Bilmes, J.: Submodularity in data subset selection and active learning. In: Proceedings of the 32nd International Conference on International Conference on Machine Learning - Volume 37. ICML'15, pp. 1954–1963, JMLR.org (2015)

Author Index

A
Altman, Eitan 90
Amini, Arash 100
Arnouss, Mohamed 167
Ashok Krishnan, K. S. 57

B
Bayiz, Yigit Ege 100
Ben Mazziane, Younes 90
Bilbao, Miren Nekane 13
Brun, O. 181
Busacca, Fabio 68
Busic, A. 25
Bušić, Ana 57, 123

C
Cardinal, Julien 123
Croce, Daniele 68

D
De Pellegrini, Francesco 90
Del Ser, Javier 13
Deugoue, Gabriel 79

F
Falco, Mariana 68
Fourneau, J. M. 25
Fu, Jie 79

G
Garnaev, Andrey 35
Getino-Petit, Mikel 13
Großmann, Gerrit 112
Gupta, Manu K. 147

H
Haddaji, O. 181
Hayel, Yezekael 135, 167

Hemachandra, N. 147

I
Ivanova, Larisa 112

K
Kamhoua, Charles 79
Kesidis, George 1
Konstantopoulos, Takis 1
Kouam, Arnold 79
Kouam, Willie 167

L
Le Cadre, Hélène 57
Le Corre, Thomas 123
Li, S. 25
Lunven, A. 25

M
Ma, Haoxiang 79
Maillé, Patrick 47
Marculescu, Radu 100
Mboulou-Moutoubi, Cleque Marlain 90
Mesabah, Islam 112
Mitra, Anirban 147
Mofouet, Romaric 79

N
Nosrati, Farzam 68

P
Palazzo, Sergio 68
Poduru, Sai Leela 112
Prabhu, B. J. 181

S
Scarvaglieri, Antonio 68
Selby, David A. 112

T
Tabrizian, Mohaddeseh 112
Topcu, Ufuk 100
Trappe, Wade 35
Tsemogne, Olivier 135
Tuffin, Bruno 47

V
Vollmer, Sebastian J. 112

Z
Zazanis, Michael 1

MIX
Papier aus verantwortungsvollen Quellen
Paper from responsible sources
FSC® C105338

If you have any concerns about our products,
you can contact us on
ProductSafety@springernature.com

In case Publisher is established outside the EU,
the EU authorized representative is:
**Springer Nature Customer Service Center GmbH
Europaplatz 3, 69115 Heidelberg, Germany**

Printed by Libri Plureos GmbH
in Hamburg, Germany